Biology of Cognition and Linguistic Analysis

Alexander Kravchenko

Biology of Cognition and Linguistic Analysis

From Non-Realist Linguistics to a Realistic Language Science

PETER LANG
Frankfurt am Main · Berlin · Bern · Bruxelles · New York · Oxford · Wien

Bibliographic Information published by the Deutsche Nationalbibliothek
The Deutsche Nationalbibliothek lists this publication in the Deutsche Nationalbibliografie; detailed bibliographic data is available in the internet at <http://www.d-nb.de>.

ISBN 978-3-631-56647-3

www.peterlang.de

"In order to understand cognition as a biological phenomenon, an observer and his role in cognition must be taken into account and explained."

Humberto Maturana

CONTENTS

Foreword 11
List of abbreviations 13

PART 1. METHODOLOGICAL ISSUES

Chapter 1. Myths linguistics lives by

1.1. The axioms of orthodox linguistics 17
1.2. The written language bias 25
1.3. Representation and linguistic signs 28
1.4. Grammar 36
1.5. Meaning and communication 42
1.6. Summary 49

Chapter 2. Toward a realistic language science

2.1. Introduction 51
2.2. First generation cognitive science 53
2.3. Second generation cognitive science 56
2.4. Biology of cognition and language 61
2.4.1. An alternative to the computational model of cognition 60
2.4.2. Explaining the biology of language 64
2.4.3. Methodological implications 70
2.5. Semiotics and cognition 73
2.5.1. Semiotics as metascience 73
2.5.2. Biosemiotics 76
2.5.3. A biocultural theory of meaning 80
2.6. Distributed language and cognition 84
2.7. Summary 92

Chapter 3. Autopoiesis and linguistic analysis

3.1. A new epistemology 95
3.2. Representations and signs revisited 98
3.2.1. Representations 98
3.2.2. Phenomenology of signs 100
3.3. Interpretation as meaning-making 102
3.3.1. Sign and meaning, or what's amiss with the semantic triangle 102
3.3.2. Signs in physical context 110
3.3.3. Ontogenetic structural coupling as the basis for meaning-making ..112
3.4. Intentionality and signification 115

3.4.1. The notion of intentionality 115
3.4.2. Communication 117
3.4.3. Reciprocal causality 119
3.5. Summary 122

Chapter 4. Speech, writing, and cognition: whence the communicative dysfunction?

4.1. The great mystification 125
4.2. Speech, thought, and writing: where rationalists go wrong 128
4.2.1. Introductory remarks 128
4.2.2. Language 129
4.2.3. Language and thought 138
4.2.4. Writing 142
4.3. The concept of literacy and the 'code' model of language 146
4.4. Speech and writing as cognitive domains 148
4.5. Types of linguistic interactions 153
4.6. The bio-socio-cognitive function of speech and writing 156
4.7. Summary 158

PART 2. THE LANGUAGING OBSERVER

Chapter 5. Grammar and cognition

5.1. Science and common sense 163
5.2. Approaching the concept of grammar 164
5.2.1. What does the term "grammar" refer to?.. 164
5.2.2. How systemic is grammar? 168
5.3. The purpose of grammar 176
5.3.1. The sign system of spoken language vs. the symbolic system of written language 176
5.3.2. The orientational function of natural language and its consequences for grammar 178
5.3.3. Grammar as knowledge structure 185
5.4. Summary 189

Chapter 6. Aspect: where Russian and English meet

6.1. Introductory remarks 193
6.2. Aspect in Russian 195
6.2.1. Boundedness and totality 195
6.2.2. The morphology of aspect 199
6.2.3. The grammatical paradigm of the Russian verb 206
6.3.4. The syntax of aspect 211

6.2.5. The meaning of aspect 217
6.3. Aspect and the participle 219
6.3.1. The morphology of participles and constraints on their use 219
6.3.2. The nature of the functional asymmetry of participles 223
6.4. Aspect in English 228
6.4.1. Divorcing tense from aspect 228
6.4.2. Aspect undisguised 233
6.4.3. Pedagogical implications 238
6.5. Summary 241

Chapter 7. Interpreting linguistic structure: how we know who does what when

7. 1. Complex sentence as a cognitive structure 243
7.1.1. Introduction 243
7.1.2. The data 245
7.1.3. The hypothesis and discussion 249
7.2. The cognitive distinction in infinitive/participle clausal arguments 254
7.2.1. The problem 254
7.2.2. What non-finite clausal arguments describe 257
7.2.3. Clausal arguments as structures for knowledge representation 260
7.3. Summary 264

Chapter 8. Nominal gender: a case from Russian

8.1. The problem with traditional semantic classification 267
8.2. Nouns vs. pronouns 270
8.2.1. The logic of naïve classification 270
8.2.2. Etymology 272
8.3. The hypothesis 273
8.4. Case study 276
8.5. Summary 280

Bibliography 283

FOREWORD

For quite some time philosophers, psychologists, sociologists, linguists, semioticians, and many cognitive scientists in general, have been speaking of a necessity to revise the starting assumptions in approaching the problem of human mind and language, which does not seem to have a solution within the traditional Cartesian paradigm. A thoroughly dualistic picture which orthodox linguistics draws of language precludes any progress in the study of human linguistic behavior as something that makes *homo sapiens* so uniquely special. The nature of linguistic ability as a most important and intriguing cognitive endowment of humans, as well as the role of language in the evolution of our species, have not yet been fully understood nor explained. Myths linguistics lives by continue to hold strong, and common sense reasoning about language, done in the very same language, is often taken for scientific explanation. No wonder linguistics as a science cannot boast of any spectacular achievements, even though it is by far the oldest among all other sciences.

Biology of Cognition and Linguistic Analysis continues the theme of an earlier book (*Sign, Meaning, Knowledge: An essay in the cognitive philosophy of language.* Peter Lang, 2003), in which it was suggested that a new epistemology was needed to overcome the methodological crisis in humanistic science. Such epistemology, it was argued, was provided by autopoiesis as the theory of the living. This book is a further attempt to show that various contemporary frameworks for the study of human cognition and language converge on the ideas of autopoiesis in that they view cognition as a biological process with a specific function, even though in such frameworks autopoietic terminology may not be used and autopoiesis itself may not be referred to explicitly or acknowledged as an underlying ideology.

In this book I explore the possibility of breaking away from the old tangle of orthodox linguistics as a non-realist enterprise in explaining what cannot be properly explained in the framework of Cartesian rationalism. Building on the epistemology of autopoiesis and integrating various approaches within contemporary non-Cartesian cognitive science into a holistic methodology, when cognition is understood as grounded in the dynamics of biological self-organization, I argue for a unified realistic language science that would view language as a biologically based, socially conditioned, cognitively motivated, circularly organized semiotic activity in a consensual domain of interactions. Linguistic analysis based on the epistemological maxim of autopoiesis "Everything said is said by an observer to another observer" facilitates deeper insights into the nature and function of language as a specific cognitive dimension of our existence.

LIST OF ABBREVIATIONS

1	First person	INF	Infinitive
2	Second person	INS	Instrumental
3	Third person	L	Locative
A	Accusative	M	Masculine
Adj	Adjective	N	Nominative
D	Dative	PA	Participle
DEF	Definite Aspect	PF	Perfective
F	Feminine	PL	Plural
FUT	Future	PR	Present
G	Genitive	PRED	Predicative
IMP	Imperfective	PT	Past
IMPERS	Impersonal	REFL	Reflexive
INDEF	Indefinite Aspect	S	Singular

PART ONE

METHODOLOGICAL ISSUES

CHAPTER 1.

MYTHS LINGUISTICS LIVES BY

1.1. The Axioms of Orthodox Linguistics

Linguistics as the scientific study of language and communication has a long history. It is probably not an exaggeration to say that the study of language not only predates all other fields of study such as philosophy, mathematics, or physics, but that it is in fact *the* primary science as language is prior to all other sciences, both historically and logically. Moreover, as Roy Harris (2005) has convincingly shown, science itself is a construct of language, held together by means of an idiosyncratic semantics designed to give it credibility: science is a construct of language because scientists impose their language on what they assume is there to be named by that language. The role of language in the lives of humans is of paramount importance, yet if we ask, "What notable effects has the 2-millenium-long study of language produced on human *poiesis* and *teoria* as distinguished from *praxis* (in the Aristotelian sense)?" the answer, for all practical purposes, will be, "None to speak of" — unlike, for example, in physics, chemistry, biology, or computer science, to name but a few.

Humans became concerned with language as an object of study when two major practical tasks emerged as a result of growing cross-cultural contacts — *foreign language acquisition* and *translation.* As Giordano Bruno is claimed to have said, "From Translation all science had its offspring" (Yates 1934: 89). However, to this day linguists have not offered an efficient methodology to deal with the first problem, and as for translation from one language to another, it still remains a fine art that escapes algorithmization. Machine translation as an aspiring project of first generation cognitive science (that is, cognitive science whose foundational concepts are based on the metaphor 'thinking is computation') continues to remain a project, and the best way to learn a foreign language, unsurpassed by any of the methods linguists might offer, is still direct cultural immersion.

Another problem that has so far resisted solution is *natural language acquisition.* As Ray Jackendoff (1994: 26) observes,

> [A]n entire community of highly trained professionals, bringing to bear years of conscious attention and sharing of information, has been unable to duplicate the feat that every normal child accomplishes by the age or ten or so, unconsciously and unaided.

Now why can't 'highly trained professionals' duplicate the child's feat of natural language acquisition? Could it be that the epistemological foundation of linguistics, and language sciences in general, comes short of explaining the very nature of language? What if 'highly trained professionals' have been trained to do something which has little to do with natural language as an empirical phenomenon and a lot to do with what they *think* language is? And what *is* language? Because we 'do things with words' (Austin 1962), we think of language as a kind of purposeful activity, and it is common belief that the purpose of such activity is to express thoughts:

> [T]he central role of language is to express thoughts. Derivatively, it has at least these two roles: explaining behavior and informing us about the world. Meanings are the properties that enable to play these roles (Devitt and Sterelny 1999: 5).

Meanings are the 'properties' of words, so the study of meaning lies at the heart of linguistic science concerned with language as a sign system 'for expressing and communicating thoughts'. To use Ray Jackendoff's (2002: 267) metaphor, meaning is the "'holy grail' not only of linguistics, but also of philosophy, psychology, and neuroscience — not to mention more distant domains such as cultural and literary theory". However, the following observation made by Jordan Zlatev is, to many, indisputably true:

> Our conception of *meaning* has become increasingly fragmented, along with much else in the increasing 'postmodernization' of our worldview. The trenches run deep between different kinds of meaning theories: mentalist, behaviorist, (neural) reductionist, (social) constructivist, functionalist, formalist, computationalist, deflationist... And they are so deep that a rational debate between the different camps seems impossible. The concept is treated not only differently but *incommensurably* within the different disciplines (Zlatev 2003: 253).

Isn't it odd that the core discipline within the field of language sciences, that is, the study of meaning, should be nothing more than a motley array of fancy theories causing feuds among their authors and followers? But couldn't then the situation be interpreted as an indication that there is simply *no* theory? And if we concede to this assumption, can we really speak of linguistics as the 'scientific study' of language?

So far linguistics has not succeeded in bringing together in a non-contradictory fashion the two concepts of language, language as a sign system allegedly used for representing knowledge ('expressing thoughts') and language as communicative activity allegedly used for 'explaining behavior and informing about the world' (that is, for 'exchanging information'). I use the word *allegedly* because these two conceptions of language reflect commonly shared beliefs that *have never been empirically validated*. For example, in communication knowledge, especially socially and culturally shared knowledge, most of the time is taken for granted, it is "presupposed and therefore *not* expressed in talk or text" (Dijk 2006: 6). We believe that linguistic signs 'represent' something such as knowledge because they have meanings, yet we don't know what meanings are, or how signs 'represent' — that is, there is no unified theory explaining both. We believe that by making inventories of linguistic signs and their meanings (dictionaries) and rules for combining them (grammars) we enable speakers of one language to learn another (foreign or second) language by using appropriate textbooks — yet those with relevant experience will argue that it is simply impossible. To make these rather mundane observations is to beg the question, "What is wrong with linguistics as a science? Could it be based on some grave conceptual fallacies? "

I am going to make a blatant claim that it is methodological inadequacies of traditional linguistics in general, and linguistic semiotics in particular, that have been to a large degree responsible for the lack of serious progress in the study of natural language as an empirical phenomenon. The problem with linguistics in general, especially its once major branch known as generative grammar, is committing a classical error of overlooking one's priors in trying to establish the place of the science of language among other empirical sciences. For example, in a recent special issue of *The Linguistic Review* dedicated to the status of linguistics as a cognitive science, claims have been made that there are solid grounds to believe that "in the past 50 years linguistics has progressively established itself as a genuinely scientific discipline" (Boeckx and Piattelli-Palmarini 2005: 447). Furthermore, it was argued that

> [S]ince language can, indeed, be profitably studied as a natural object, the study of language should share the developmental paths, the assumptions and the explanatory style of the most successful natural sciences, epitomized by theoretical physics (*ibid.*, p. 462).

Somehow it escaped the authors' notice that even though the Chomskian revolution in linguistics in the 1950s did in fact turn linguistics into a branch of cognitive science, it did not make generative grammar the one and only legitimate framework for the study of language. Cognitive science itself has since far outgrown generativist assumptions about the nature of cognition in general and language in particular, positing new foundational epistemological principles and using novel approaches in the empirical study of human cognitive abilities (more on that later). There has been much talk among generativists about the role of linguistics in cognitive science whose ultimate objective is to understand how the mind works (Ritter 2005) or, as Zenon Pylyshin (1999) puts it, to understand "what's in our mind". However, the infamous generativist claims that 'thinking is computation' and the mind is a 'syntactic engine' for carrying out such computation, that language 'is in the mind' (Culicover 2005) and, therefore, a property of the mind, a kind of mental faculty or organ (Anderson and Lightfoot 2002) that grows naturally in the biologically normal individual on the basis of limited external linguistic input, encompassing computational strategies and specific contents, can hardly be taken seriously any more. As for the call for linguistics to follow 'the developmental paths, the assumptions and the explanatory style of the most successful natural sciences, epitomized by theoretical physics', it sounds both misplaced and outdated. Misplaced, because if theoretical physics is best represented by quantum physics, then the explanatory style of the latter adopted by language sciences should inevitably lead to the conclusion that nothing is determinate (Aarts 2007), Principles and Parameters or not, and everything is at the same time something else — while this is exactly what generativists have always argued against as ludicrous — at least, on their view. Outdated because, as has been mentioned at the beginning, contemporary serious research into the nature and function of human knowledge (the purported goal of philosophy and natural sciences that sprang from it) starts with an assumption that language is prior to science. As observed by Antoine Lavoisier (1789, 3d paragraph),

> It is impossible to disassociate language from science or science from language, because every natural science always involves three things: the sequence of phenomena on which the science is based; the abstract concepts which call these phenomena to mind; and the words in which the concepts are expressed. To call forth a concept a word is needed; to portray a phenomenon a concept is needed. All three mirror one and the same reality.

Linguistics simply cannot follow the developmental paths of natural sciences because these themselves developed with the development of language, particularly its written form which was brought to life, among other things, for the specific purpose of serving science in its quest for knowledge about the world.

Now, what is the problem with linguistic semiotics? Essentially, it is that for a long time the true nature of linguistic signs was obscured by strong beliefs that signs were artificial, conventional, and arbitrary entities intentionally produced by humans for the purpose of communication understood as exchange of encoded meanings. These encoded meanings (or 'mental content', 'thoughts', 'abstract knowledge', 'complex propositions', 'conceptual structure', etc.) are assigned to special material entities known as 'sign forms' or 'sign vehicles' in order to be conveyed from the speaker's head into the listener's head via the medium of oral speech — hence the so-called 'code-model' of communication based on the 'conduit' metaphor (Reddy 1979). Communication in this context becomes something superficial and irrelevant (as in generativism), and as a result linguistic behavior (*languaging*) is abstracted from the physical context of its occurrence in real time and space. It is, therefore, not surprising that both generative linguistics and classical representational theories of meaning have little to offer in the way of explaining real living language. To quote Roy Harris (2004: 735), "the 'circumstantial' parameter of human communication... has been scandalously neglected in the history of Western linguistics". Talbot Taylor (1997) sides with Harris in criticizing language theory for removing language from what actually gives it its life:

> We have today an academically enshrined linguistic science which takes as its data a decontextualized, a historical, and autonomous product, ignoring the voluntary, contextualized actions of individual agents in producing that data (p. 149).

Moreover, Taylor demonstrates that what orthodox linguistics normally excludes from analysis as something 'external', by drawing a divide between *linguistic* and *extralinguistic,* such as use, practice, history, etc., is in fact a very important *internal* characteristic of language. Thus the truly essential properties of language as a physically grounded biologically, socially and culturally determined joint activity of humans remain largely ignored.

On the one hand, it is very true about language what Cornish-Bowden et al. (2004: 715) have said about biology:

> In practice, most current research <...> is not directed towards solving fundamental problems, but towards accumulating more detailed facts, and unfortunately this may cloud understanding as much as it improves it.

On the other hand, there is a growing realization among language scientists that until recently linguistics has been busy not so much with the study of language as a natural phenomenon, but of a certain set of theoretical constructs built on a set of often intuitive and empirically unvalidated assumptions and premises. As a result, instead of the study of language as a kind of patterned cognitive activity characterized by complex dynamics, linguistics has largely become the study of theoretical models and conceptual frameworks most of which, one way or the other, lean heavily on the axioms of orthodox thinking about language. Moreover, just because habit often becomes our second nature, the mere fact that we have become used to something, that we have grown up with certain beliefs instilled in us over the years of schooling and indoctrination, usually prevents us from feeling even the slightest temptation to question the so-called 'well-known truths'. And much of the theoretical inheritance of traditional linguistic science is just such a compendium of well-known truths, and to even question them is equal to committing sacrilege by desecrating something made holy by time and tradition. But, as Nikolai Zhinkin (1998: 80) warned,

> [M]ost dangerous is the case when an interpretation established by tradition has a subconscious effect such that a man passes by the facts without even wanting to pay attention.

Because within the traditional paradigm the empirical essence of natural language eludes a clear-cut definition, the ideal project of linguistics as a science remains, strictly speaking, unspecified: what is the ultimate goal linguistic science sets out to achieve that, in fact, makes linguistics a science? Three questions seem to be invoked here, namely, *what language is*, *what language does* for us, and *how it does it*, and depending on how we answer the first one we may be closer to, or farther away from, illuminated answers to the other two. Seeking to answer these questions, orthodox linguistics[1] builds its grand palace on the following seemingly obvious but, nevertheless, *intuitive assumptions about language and cognition* which, overtly or covertly, imply that humans are input-output systems:

1. Language is something 'out there', in the real world, and an important stage in the development of every human individual is 'acquisition' of language.
2. Language is a system of material signs (a code) in which every sign form expresses a certain meaning (or a set of related meanings) attached to it.
3. Linguistic signs represent in the mind objects and things that exist independently in the real world, and our thoughts about these objects: "To speak a language is to re-present (to present again) the perceived world by selecting <...> some of its characteristics" (López-García 2005: 72).
4. The meanings of linguistic signs are given in advance and may be identified and described as abstract invariants on which humans draw in communication as needed.
5. Since knowledge is the reflection in the mind of aspects of reality which itself is represented by linguistic signs, language is a symbolic system for knowledge representation.
6. The function of language is to transfer meanings (thoughts or mental content) from head to head, that is "telementation".
7. The transfer of thoughts by means of language lies at the core of human communication.
8. Communication is exchange of information (knowledge) between the sender and the receiver, and the system of linguistic signs is the conduit for such exchange.

---

[1] By 'orthodox' I refer to linguistics whose theoretical assumptions about language and communication are epitomized in encyclopedias, dictionaries, school and college texts and the like.

These basic assumptions underlie specific propositions about particular objects of study which characterize approaches taken within various linguistic fields — such as *linguistic semiotics* (assumptions 2 and 3), *linguistic semantics* and *pragmatics* (assumptions 4 - 6), *computational linguistics* (assumptions 2 - 6), and *communication and discourse studies* (assumptions 6 - 8). Various theories developed in the frameworks of specific linguistic disciplines in orthodox science bear the birthmark of these initial axiomatic assumptions and speak unequivocally of their methodological foundation which is *Cartesian dualistic philosophy.* Orthodox linguistics is permeated with dualism throughout, and regardless of its fleeting infatuations with one methodological trend or another over the past hundred years or so, it has shown remarkable tenacity in sustaining the two dualistic assumptions about language and communication as its cornerstones: (1) language is a code and its function is that of representation, (2) communication is telementation. Linguistics lives by these metaphors failing to see that they are nothing but myths, thus falling victim to grave conceptual fallacies.

Many researchers in the language sciences have been increasingly dissatisfied with traditional views on language and cognition. Recently, there have been a lot of criticisms of the core assumptions of orthodox linguistics (Taylor 1992, 1997; Lamb 2004; Love 2004; Millikan 2005 *inter alia*) and mainstream cognitive science. Cognitivism as it was originally conceived by the 'founding fathers' at the dawn of the computer era, postulates that certain behaviors (e.g. language) can be explained only by appeal to 'internal cognitive processes'. These internal processes consist of mental representations of information and a finite set of rules that operate on them. The mind's capacity to automatically receive, store, manipulate, and output INFORMATION depends on its receiving, storing, manipulating and outputting REPRESENTATIONS of that information (Eckardt 1993). On the internalist account of cognition, cognitive processes occur within the skin (Harnad and Dror 2006). By accepting an internalist account of mind and language, first generation cognitive science fails to see the important methodological consequence of separating, or in Harris's (2004) terms, 'segregating' language and mind from the so-called 'external' physical world. As argued by Eleanor Rosch (1999), mind and world occur together in a succession of situations which are somewhat lawful and predictable and, as pointed out by McGee (2005a), experimental results challenge certain notions about the existence of an objective, timeless reality.

New ideologies and approaches within cognitive science have emerged, such as *embodied mind* (Varela, Thompson, Rosch 1991; Lakoff and Johnson 1999) and its 'next-of-kin' *extended mind* (A. Clark 1997; Clark and Chalmers 1998), *radical constructivism* (Glasersfeld 1995), *dialogism* (Linell 1998; Linell, 2007; Thibault 2000; Hodges, 2007), *integrationism* (Harris 1996, 2004; Cowley 2002; Sutton 2004), *distributed cognition/language* (Cowley 2004) — to name but a few. Thus we can speak of 'second generation cognitive science' (Lakoff and Johnson 1999) or *enactive cognitive science* (McGee 2005a, 2005b). It views perception as consisting in perceptually guided action, and cognition as not the representation of a pregiven world by a pregiven mind but rather as the enactment of a world and a mind on the basis of a history of the variety of actions that a being in the world performs. In Ernst von Glasersfeld's (1995: 51) words, "cognition serves the subject's organization of the experiential world, not the discovery of an objective ontological reality".

Whatever the differences in the stances taken within each particular framework, they all share what seems to be a given, that is, the biological, or bio-socio-cultural nature of language and cognition. Thus biologism appears to be the new general ideology with a very strong appeal to cognitive scientists (Miller and Lenneberg 1978; Millikan 1984; Plotkin 1987; Givón 2002, Kravchenko 2006 *inter alia*). This naturally raises the issue of new epistemological foundations for cognitive science as Cartesian philosophy appears to be strikingly at odds with the biological approach, precluding any meaningful discussion of the vast empirical data accumulated over the years by researchers in the field of cognitive science. As has been argued elsewhere (Kravchenko 2003a), such foundations have been laid out by the *biological theory of cognition* or *autopoiesis* (Maturana 1970, 1978; Maturana and Varela 1980, 1987) as the theory of organization of the living. The epistemology of autopoiesis allows to overcome the shortcomings of traditional philosophy of mind based on Cartesian logic with its ontological distinction between mind and body (Priest 1991; Schlechtman 1997; Kim 1998 *inter alia*). The 'tectonic' biological shift in contemporary cognitive science — and, particularly, in cognitive linguistics — indicates that the time has come to integrate linguistic research with autopoietic epistemology. Such integration may lead to deeper insights into the nature of language and its function, relating it to the adaptive function of humans realized in the domain of cognitive/linguistic interactions with the environment.

The aim of this book is twofold, so it consists of two parts: (1) "Methodological Issues", and (2) "The Languaging Observer". In the first part, I will argue for the need to re-evaluate if not all, then at least some of the common assumptions about language and communication mentioned above. I will outline the direction taken by language sciences toward integration on a basis of the biological theory of cognition which views language and cognition as adaptive activities of structure determined organisms (living systems) in a consensual domain of interactions. In the second part, I will show the empirical reality and linguistic utility of the central autopoietic thesis about language, "Everything said is said by an observer to another observer who can be himself or herself", by providing some case studies for analyzing particular linguistic phenomena in two typologically different languages, English and Russian. These case studies are intended to illustrate that the theory of language advocated for in the first part fully meets the standard criteria for theory adequacy formulated by Kuhn (1977) — accuracy, consistency, scope, simplicity, and fruitfulness.

1.2. The Written Language Bias

The understanding of language as a code has been so deeply ingrained in the metalanguage of linguistics that the ubiquitous code model has become part and parcel of any theoretical construal pertaining to language — from linguistic semiotics to discourse analysis. In fact, this model underlies almost all theorizing in cognitive science and cognitive linguistics as well. However, there have always been voices (Ferdinand de Saussure no exception) warning linguists against confusing language as a *natural phenomenon* (a kind of patterned behavior) which should be the true object of linguistic study, with language as a *system of written signs* designed to represent (very often in a quite arbitrary fashion) short-lived acoustic phenomena that accompany communicative behavior. Although written language is in fact a system based on the use of a limited set of characters (such as letters in alphabetic systems of writing) and, because of this, resembling (at least, superficially) a code, it does not *represent* language as a specific kind of cognitive interactions with the environment. Spoken and written language possess different ontologies and simply cannot be viewed as two different manifestations of the same phenomenon.

Over the observed history, linguistics has been misled in its endeavors by what may be called the 'great mystification' pertaining to its original legitimate object of study, that is, natural (spoken) language. Surely, the role of written language today is understood not to be limited exclusively to 'representation' of spoken language, and although written language develops later in cultural history and in language epigenesis, Per Linell (2005) is surely right when he emphasizes that "it is *not* true that written language is simply secondary to spoken language" (p. 26; see also: Harris 1995; Pettersson 1996; Hagège 1996). Writing is used simply for different purposes than speech, speech and writing have a considerable amount of mutual autonomy and independence. Preoccupation of linguists with the study of cultural artifacts such as written signs precludes any sensible talk of linguistics as a study of natural language viewed as a biological endowment of humans with a specific function. According to Talbot Taylor (1997), traditional linguistic theories are not only misleading, they are wrong because they misrepresent their object of analysis.

Counter to orthodox views on language as a means of exchanging information (communication), the primary function of spoken language is to exert, in the course of interactions with others, orientational influences on human behavior in real time-space (Maturana 1978; Bottineau, in press), while the primary function of written language is to serve as the external analog of internal, or biological, memory, as "a storage and retrieval system that allows humans to accumulate experience and knowledge" (Donald 1991: 309). The evolutionary purport of writing is of a different nature than in the case of spoken language: writing helps overcome the spatio-temporal limitations of spoken language by establishing an atemporal universal (as contrasted with the temporal and local, in the case of the oral tradition) existential/informational environmental medium for humans. This informational medium is akin to the human genome in biology, since it is only in and through this medium that we become humans. Yet, as Per Linell (2005: 30) emphasizes,

> we can talk about a *paradox* in modern linguistics: one claims the absolute primacy of spoken language, yet one goes on building theories and methods on ideas and experiences of a regimented, partly made-up language designed for literate purposes and overlaid with norms proposed by language cultivators, standardizers and pedagogues.

Herbert Clark (1996: 24) notes that "it makes no sense to adopt an approach to language use that cannot account for face-to-face conversation, yet many theorists appear to have done just this". As a result, many so-called 'truths' about particular linguistic phenomena which have been 'discovered' as a result of linguistic analysis of written language, have little, if anything, to do with the real facts of spoken language. Suffice it to mention the notorious double negation rule imposed on English grammar by the bishop Robert Lowth (1762/1767). This rule, despite strict insistence on the part of English language teachers, continues to be defied in everyday use by native speakers, and triple and even quadruple negation is not uncommon in the speech of educated people, cf.:

(1) *I ain't tellin' nobody nothin' no way.*

Or take the conviction that the primary function of personal pronouns is anaphoric, although, as Huddleston (1984) convincingly argued, we all use them indexically without impeding understanding in any way. Or that verbal aspect has to do with the internal temporal structure of events (more on this in Chapter 6), or that the difference between the demonstratives such as *this* and *that* is one of distance between the speaker and the referent, etc. — the list of similar misconceptions may be continued.

In his very thorough and fascinating analysis of the written language bias, its nature, origins and transformations, Per Linell (2005) has shown how the great mystification of linguistics came about, what historical and cultural developments helped reinforce the false belief that writing represented speech and, above all, what makes written language so radically different from spoken language. The most important distinction between speech and writing is the relationship between language production, perception and interpretation. Speech is *dynamic behavior distributed in real time,* and its inherent dynamics must be subject to on-line monitoring and analysis by *both communicating parties.* The products of the speaker's activities, that is the behavioral movements and the resulting sound waves, are ephemeral; they fade away rapidly, and the same applies to the listener's activities. As a consequence, humans engaged in communication are forced to focus on the dynamic behavior as such, rather than on some persistent products, like in writing. Talk-in-interaction involves bodily gestures; participants use their voices, faces, bodily orientations and movements to signal messages. In brief, spoken language and interaction are *embodied.* At the same time, the whole interaction between speaker and listener is *interdependent with the situation,* and other contexts, in many important ways. The situation *as perceived by* participants will change as the interaction moves ahead.

Now, what we call written language, Per Linell argues, is different from spoken language in many ways. Unlike speech, written texts are typically *not* perceived and interpreted at the same times and places as they are produced. The analysis of written language necessarily focuses on the *products* of the writer's activities, on the written texts, whereas the production process itself is inaccessible and unimportant for the normal reader. A written text and its component parts (letters, words, sentences, paragraphs, etc.) have the character of *objects.* They are *persistent* and *static,* and are spatially rather than temporally organized. A written text is made up of *discrete* symbols, i.e. letters and graphic words, which are organized in certain regular *spatial* patterns. These symbols are only very approximate counterparts of some of the structural features of spoken language (such as segmental-phonological, grammatical, lexical); as for the prosodic features and the non-verbal signals of the communicative acts, they have almost no correspondence in writing. The written text exists as a symbolic artifact, which is not embodied by its users, the communicating people. Importantly, written language is *disembodied* with respect to the human agent.

Embodiment and dependence on the situation (physical context in real time) are the two dividing lines between spoken and written language making them empirically non-identical and non-equivalent. Although, because of the written language bias in linguistics, language is conceived of as a kind of code, when it is assumed that there is one-to-one correspondence between the sign (e.g., the word) and what it stands for, it is a well known fact that no such correspondence is observed either in natural spoken language (communication) characterized by a large number of situational (contextual) variables affecting the actual outcome of communication as a kind of joint activity (Love 2004; Farrow 2005), or in written language which seldom allows single unified interpretations of particular pieces of writing. Historically, this multiple interpretability of texts motivated the emergence of *hermeneutics* as the science of interpretation.

The essential ontological difference between spoken and written language is the difference between first order activity, that is, doing what you talk about when you talk about action and perception (having both a neural and behavioural aspect), and second order activity involving operations on (second order) cultural constructs (words, phrases, sentences, ets.) that make up the 'ultimate' artifact, written language. Just in the same way as verbal descriptions of the environment, no matter how thorough and finely detailed, cannot make it present to the blind person the way eyesight makes it present, a written 'representation' of spoken language cannot make present all the fine and versatile details of the actual utterance-in-situation that our hearing, *along with all the other senses*, does. And the important thing to understand is, of course, that the function of written language, initially, is *not related to representation of spoken language*. The idea that written language represents spoken language springs from the time-old belief (let us call it the 'original representation axiom') that words of spoken language represent objects in the world, that physical words as acoustic phenomena are special *signs* that stand in for something not necessarily visible or otherwise perceptible to the senses. Likewise, by an obviously trite analogy, written words as visually perceived graphic patterns (something 'present') begin to be viewed as signs of acoustic phenomena which, for one reason or the other, are inaccessible to auditory perception (that is, they are 'absent'). However, our everyday linguistic experience tells us (if we choose to listen, of course — which, typically, doesn't happen very often) that interpretation of written expressions as if they were representations of spoken expressions often creates cognitive problems which affect our ability to choose adaptively appropriate behavior in a particular situation. The effect of the written language bias on both the interpretation and, strangely enough, production of written expressions in different cognitive domains of interactions, will be discussed in more detail in Chapter 4. Now we'll take a closer look at the 'representation axiom', that is, that linguistic signs represent in the mind objects and things that exist independently in the real world, and our thoughts about these objects.

1.3. Representation and Linguistic Signs

The concept of (mental) representation lies at the core of any theory of signs, but anyone with an interest in semiotic issues (both philosophical and linguistic) knows only far too well that the concept of mental representation as used in contemporary literature is very fuzzy and elusive and its more or less consistent use unavoidably invokes one question that has to be answered prior to any productive discussion of the nature of cognition and cognitive capacities: *What is a mental representation?* As observed by Wells (2000: 116),

> it is generally agreed that human mental activity has to do with representations but, beyond that, there is considerable difference of opinion on such questions as: Do other species also make representations? If so, for what purpose? What is the nature of representations? Are they stored in the mind and retrieved in a relatively fixed form, or are they created on the fly in response to the demands of the situation? How are representations related to propositions realized in linguistic form?

In contemporary philosophical theory of knowledge representations are understood as certain mental structures including intentional categories *(believe that p, wish that q)* which constitute the content of linguistic (semantic) structures at the deep level. In psychology, representations are typically described as conceptual structures individuated by their contents (Margolis and Laurence 1999) and defined in accordance with the traditional methods of analytic philosophy, that is, by positing sets of necessary and sufficient conditions that have to be met. However, the notion of conceptual structure widely used in contemporary cognitive science remains, largely, a 'thing-unto-itself': although many authors write about it as something that unquestionably exists in the mind, the accepted practice doesn't make it any more lucid for the simple reason that mind itself (its nature and function) remains as yet an unresolved problem.

Despite a long history of discussion (see, e. g., Watson 1995) it has become obvious that there is no way out of the philosophical dead-end within the framework of the old rationalist paradigm since, as argued by Stephen Stich (1992), traditional analytic philosophy is incapable of providing a reasonably consistent and empirically sound unified theory of mental representations. There have been attempts to re-evaluate the notion of representation without departing from the mainstream approach. For example, in his endeavor to outline, apply, and defend a theory of natural representation, Rick Grush introduced the "off-line" computational metaphor to describe representation as opposed to presentation which is accessed in the "on-line" mode:

> [W]hat distinguishes presentations from representations is the use they are put to. A presentation is used to provide information about some other, probably external in some sense, state of affairs. It can be used in this way because the presentation is typically causally or informationally linked to the target in some way. The representation's use is quite different: it is used as a counter-factual presentation. It is, in very rough terms, a model of the target which is used off-line to try out possible actions, so that their likely consequences can be assessed without having to actually try those actions or suffer those consequences. Second, and this is implicit in the first point, the ability to use an entity as an off-line stand-in depends crucially

on its not being causally linked to, and its not necessarily carrying information about, the entity it represents (Grush 1997: 7).

A similar approach is taken by Chris Viger (2007) who distinguishes between two notions of mental representation: *presentations*, which he defines as 'stimulus bound', as information from something transduced by our sensory systems so that we can modify our behavior, and *symbols* as mental representations which stand in for something that is not present. Although Viger makes a disclaimer saying that he does not agree that starting with cases of directly sensing something is the way to proceed in developing a theory of mental content, he does, however, admit that there is something very compelling about the view. This view, he argues,

> seems to follow from what is surely true, which is that very often if I am staring at a tree in good lighting conditions that explains why my thoughts are about the tree. What does follow from that truism is that a tree very often tokens TREE in me; however, what needs explaining is why that mental representation has the content *treehood* for me (Viger 2007: 127).

It might be pointed out at once that the mere fact that I am staring at a tree — that is, that the direction of my gaze can be described using this particular linguistic expression — does not necessarily imply that I am *thinking* about the tree, nor that I have any coherent thoughts about whatever at all. 'To stare at something' means to have one's head and eyes spatially oriented so that a straight line could be drawn between the pupils of one's eyes and that something one is said to be staring at, but it does not mean that because such a line can be drawn the person who is staring *must* have thoughts about whatever is at the other end of the line. But this is entirely beside the point. What is important in the context of the problem of mental representation is that both Grush and Viger seem to believe that visual interaction between an organism and the environment implies a physical interaction — hence the notion of 'trunsduction' of external information via sensory stimuli and its subsequent internal 'processing' characteristic of common sense psychology as well as many scientific theories about human behavior. However, as argued by Timo Järvilehto (1998), the notion of 'trunsduction' is based on the separation of man and environment and is, to put it mildly, inadequate, and to put it strongly, outright false. In modern psychological theories, in particular in cognitive psychology, humans are viewed as input-output systems when mental activity is based on linear processing of information from the environment: sensations, perceptions, and mental activity in general, are regarded as something related to the inner processing of environmental information by the senses and by the brain. The senses receive the stimuli which then run along the nerve paths to the centers analyzing the incoming information and creating interpretations of the events occurring outside the brain and body. This is the internalist account of mind, and it has shown itself to be much less productive (if at all) than was originally expected — even for the reason that there is no information pre-existing in the world (Reeke and Edelman 1988), and a living organism is a structure determined system, that is, a system "such that all that happens in it or with it arises as a consequence of its structural dynamics, and in which nothing external to it can specify what happens in it, but only triggers a change in its structure, determined by its structure" (Maturana 2000: 461). Structural determinism provides an epistemological basis for a better honed biological theory of perception when

> in any visual interaction the cells in fact respond to the relations in which the physical parameters are given. This can occur only because, although for us, as observers, relations are abstract notions, the relations in the physical parameters to which the cells respond are embodied in the anatomical and functional organization of the system. The anatomical and functional organization or the functional states constitute an embodiment of relations that determine the course of the future interactions of the system. Thus, *although in the strict physical sense all visual interactions are physical interactions, in the biological sense they are abstract interactions in which it is the relations of the parameters of light, and not their physical properties, that determine the response of the system* (Maturana, Uribe, and Frenk 1968: 29)

Commenting on this, Seiichi Imoto (2004a: 13) emphasizes that

> the visual field is not a physical stimulus field; it is always and already composed of the constructs that were made through the visual processes by "the relations of the parameters of light". In this sense, what we see in front of us, for example, is neither a stimulus nor a cause for our vision, but an end result of our visual processes.

On this approach, the notion of presentation as suggested by Grush and Viger lacks the necessary explanatory power and becomes, therefore, void. As for counterfactuality (Grush) and symbolic nature (Viger) of mental representations as opposed to presentations, these appear to be alternate descriptions of representations as *semiotic phenomena* (in the traditional sense of the term, when signs are defined as arbitrary abstract symbols that stand in for something perceptually absent), thus falling within the traditional theory of mental representations. On a closer look, it seems that the entire traditional framework for the analysis of representation is conceptually misleading as it implicitly builds its construals on the assumption that representations, in virtue of being semiotic phenomena, *have to do with codified equivalence and identity*, thus overlooking the simple fact that, conceptually, in order to be codified (in the usual linguistic sense of the word), a phenomenon must first be grasped by a cognitive structure *which cannot be linguistic by definition*. Attempts to resolve the representational problem on purely linguistic grounds are limited to introspection but, as argued by William Croft (1998: 152),

> introspective linguistic evidence does not determine mental representation to as high a degree as appears to be commonly assumed. In other words, the "right" model of mental representation cannot be determined by introspective linguistic evidence alone...

Picking up the discussion, Dominiek Sandra (1998) insists that it is beyond the power of linguistics to prove anything about mental representation at all. Thus, if we consider an entity such as a specifically organized sequence of sounds categorized as a word (e.g. *smoke*), there does not seem to be any ontological dependency between this entity and that entity of which the word *smoke* is believed to be a sign. For a newly born baby linguistic signs that the baby begins to interact with *are not signs* (that is, something perceptible that stands in for something perceptually absent), just like any other kinds of entities constitutive of the baby's environmental niche are not signs, either. If we consider the possible orientational effects which various environmental components may exert on a baby, the sound of a thing accidentally hitting the floor will not be essentially different from the sound of a word uttered by someone, although their respective representations as specific mental 'states' will, undoubtedly, differ. There isn't, therefore, any obvious relationship between the designator (manifest entity) and the designatum (non-manifest entity). It is this simple observation on which the arbitrariness-of-sign thesis (Saussure 1922) has rested and of which linguistic science is so reluctant to let go. However, and this is also something quite obvious, it is possible to infer that signs are arbitrary only when an important epistemic maxim is ignored, namely, the maxim of *primacy of phenomenology* in cognition. Consequently, this inference cannot be but essentially false. Any entity, whether linguistic or non-linguistic, is conceptualized, categorized, and identified as such only in the course, and as a result of, an organism's interactions with the environment. According to Walter Freeman (1995), what an organism 'knows' it knows as a result of its own actions, which are uniquely situated in space and time. What we have become so much used to calling linguistic signs, opposing them to non-linguistic signs, for an observer are just another variety of components of the immediate environment (environmental niche) with which the observer may interact just like with any other entity.

When, over time, interaction with the niche constituents of the same kind has occurred a certain number of times, the corresponding mental state (representation) acquires a more or less stable (patterned) character, it becomes, for an observing self, a dynamic neuronal structure which may be described as standing in a causal relation to the niche constituents of the same type, or the prototype. After a causal relation between two entities (such as smoke and the acoustic word *smoke*) has been established, it takes the form of a complex representation incorporating elementary representations of these entities. One entity's enaction activates a corresponding elementary representation which induces an active state of another elementary representation. In such cases we say that one (ostensive) entity is the sign of another (non-ostensive) entity. But, and this is important, there is no law that predetermines the sequence of an organism's interactions with the environmental constituents in such a way that interaction with a linguistic entity necessarily either precedes or follows interaction with a non-linguistic entity. By inference, *just like a word may function as a sign of some entity, an entity may function as a sign of the word,* that is, a semiotic relationship rests on *reciprocal causality.*

This inference unavoidably follows from the fact that signs constituting natural language are nothing but empirical objects included into an organism's interactional domain on exactly the same grounds as all other non-linguistic objects, namely, on the grounds of their ostensibility/perceptibility. Yet in linguistic philosophy (particularly, in semiotics) there is a tradition to treat linguistic objects as *a priori* ostensive entities whereas all non-linguistic objects that display a causal relationship with linguistic objects are treated as *a priori* non-ostensive entities. This epistemological premise is the cornerstone of all current theories of signs (cf. Keller 1998). But how adequate is it?

If we try to conjecture an organism's domain of interactions (its niche) as some physical space constituted by a set of perceptible objects, the proportion of linguistic objects found in it may turn out to be negligibly small if not zero. A human individual may be in a situation devoid of any acoustic or visual linguistic objects, but full of different objects of non-linguistic nature. What happens in the course of this individual's perception of, and cognitive interaction with, these objects?

First, all perceived objects are categorized as recognizable and unrecognizable. As argued by Michael Leyton (1999), perception, fundamentally, is a form of *memory* which is defined as a physical object that is in the present of some observer and that has features that the observer causally explains. In a trivial perceptual situation the number of unrecognizable objects would be far less in numbers compared to the objects that can be recognized, or there may be even none at all. A recognizable object is an object which, when interactionally involved, activates a complex mental experiential/mnemonic structure (a relative state of neuronal activity), or representation. Constitutive of this complex representation is an elementary representation of a preceding interaction with this object, or a set of such elementary representations. Normally (that is, in the case of humans in good physical and mental health whose biological and socio-cultural development took place under conditions defined as natural for the genus *homo sapiens*), this set would include a representation of interactions with another, linguistic object, which stands in a causal relation to the perceived object. Both linguistic and non-linguistic objects (that is, their percepts) contribute to the formation of a single mental structure (concept) on equal epistemic grounds, that is, through experience which results in specific states of neuronal activity within the nervous system. Depending on which parts of the concept and to what degree of intensity are activated during object perception, the outcome configuration may comprise a representation of one or another linguistic object (such as a word) as a possible name for the perceived object. An important role in the process is played both by the characteristics of the perceptual situation (the totality of factors bearing on it directly or indirectly) and *the observer's individual experience*. Conversely, an unrecognizable object is an object which, when involved in interactional activity resulting in a specifically configured complex representation, does not make it possible for this particular configuration to "turn on" an existing mental structure comprising a representation of an interaction with a linguistic object, drawing it into a common conceptual structure. As a result, the problem of naming (conceptually identifying) the object may arise.

Second, in the conjectured situation just described the linguistic object has the status of non-ostensive entity because the observer who perceives, identifies, and names the ostensive entity moves from the non-linguistic object to the linguistic object. This means that the perceived (ostensive) non-linguistic object enters a semiotic relationship with the linguistic (non-ostensive) object, so we can speak of *the semiotic multiplication of the world*. It is important to note that such multiplication stands in direct relation to the measure of empirical experience of the world an observer has, including, as a default condition, his broad experience of linguistic objects. Semiotic multiplication of the world leads to an exponential increase of possible interpretations of sign relations; consequently, the capability of 'processing' categorized cognitive experience 'embodied' in signs in a non-contradictory manner becomes of paramount importance, making linguistic communication effective in the long run.

And, finally, third: since to the observer linguistic signs are empirical objects, perceived interactions between signs may also result in an emergence of causal relations. This assertion is relevant for understanding the genesis of linguistic competence (understood here in the generative sense as an abstract idealization of a native speaker's knowledge of the grammar) as it allows for a new approach to the cognitive origins of natural language grammar.

The structuralist tradition to treat language as an autonomous self-sufficient system could not but reflect on the treatment of linguistic signs which are believed to be a special kind of entities whose ontological properties are radically different from those of non-linguistic entities. Among these properties one is considered to be crucial for the definition of sign — the so-called "producibleness" of signs since they are, allegedly, the product of humans' intentional creative activity aimed at making communication possible. In orthodox science, communication continues to be understood as exchange of coded information, and despite the development of inferential theories of communication (cf. Grice 1989), when it comes to the study of language, as Deidre Wilson (1997) put it, "the code model still has us in its grip".

Indeed, such exchange (if we subscribe to the traditional view) is possible not only and exclusively through the medium of language, but through a variety of other means, including different bodily functions such as non-linguistic acoustic phenomena, gestures, facial expressions, smells, etc. which can hardly be characterized as signs of a code. When the function of language is identified as communication, the following relevant circumstance seems to be overlooked: communication as a domain of cognitive interactions of organisms (see Chapter 3) constitutes a basis for linguistic behavior, but it is not limited to it. Although this trivial fact has not been generally questioned (which is quite understandable, for were it otherwise, the following inference would inevitably follow: besides *homo sapiens*, no other biological species is capable of communication), every time the function of language is brought up for discussion a subtle metamorphosis takes place, and *communication* is identified with *languaging*: "Language, in its developed literary and scientific functions, is an instrument of thought and of the communication of thought" (Malinowski 1927: 297). Such understanding of the function of language prevails in contemporary linguistic literature (cf. Van Valen and LaPolla 1997, Fauconnier 1999). Yet this definition does not cover all the essential properties of language. According to Andrei Zvegintsev (1996: 50),

> language is an activity that involves all the functions which make humans human. And language is an activity that generates the means for its realization in concord with the diverse functions possessed by language. <...> To limit the study of language to the study of its use as a means of communication and thought is to deliberately narrow the scope of one's research and forsake cognizance of the true nature of language in its entirety.

Linguistic behavior (languaging) is just one among many various kinds of human interactions (joint activity) that may be described as communication, albeit the most effective one in terms of its ability to exert orientational influences on interacting individuals. These other means of communication may be viewed as intentional activities that include producible signs (screams, gestures, etc.); however, it does not make them kinds of linguistic behavior. So 'producibleness' as a characteristic of signs is not crucial in the definition of their ontological properties. Another consideration against viewing this characteristic as decisive is that 'producibleness' as a property implies 'createdness' of the entity this property qualifies. In regard to language, 'producibleness' implies 'creativity' as a specific feature of linguistic activity — not quite in the Chomskian sense, but in the sense that signs (words) are viewed as *artifacts intentionally created for the purpose of communication.* Such implication is admissible only and if the issue of the origin of language has been satisfactorily resolved. As long as it has not been done, postulating 'createdness' as an essential property of signs seems to be methodologically inappropriate. On the contrary, for humans acquiring language in a natural way the intentionality of 'created' (by who? when? how?) signs constituting language as a system, is not an issue at all. For humans as living organisms, from the very first moments of their lives, linguistic signs are natural components of the environment just like any other kinds of entities they begin to interact with, starting an inevitably long but indispensable process of acquiring and accumulating experience. As long as a baby / child / adolescent / adult is part of human society, accumulation of linguistic experience continues, although the speed and intensity of the process change as it goes through different stages. We shall return to the issue of phenomenology of signs in Chapter 3.

1.4. Grammar

"What is grammar?", asks David Allerton (1990: 68) in "An Encyclopaedia of Language" edited by Neville Collinge, and gives the following answer:

> A language is basically a system of signs, i.e. of institutionalised sensory patterns that 'stand for' something beyond themselves, so that they 'mean' something. Linguistic signs are arbitrary sound patterns (or, in the case of written language, visual patterns) which have a particular meaning in the language in question. ... The word 'grammar' when applied to the study of these patterns is used in two slightly different ways: whereas 'a grammar' may cover a language in all its aspects, 'grammar' (without any article) covers only part of it.

So grammar appears to be the study of linguistic signs which are 'arbitrary sound patterns' in spoken language or 'visual patterns' in written language. Definitions of this kind are a good example of basic inconsistency in how grammar is understood, this inconsistency stemming from the written language bias which leads to confusing spoken and visual patterns as basically *equivalent.* Nevertheless, if we choose to be consistent in the use of special terminology pertaining to the study of language, we cannot make light of the fact that the literal meaning of *grammar* (from Gr. *grámma* 'letter' + *-ar* 'pertaining to') is 'the art of writing'. Strictly speaking, grammar as the study of linguistic signs was from the very start a study of *written signs,* that is, sentences which made up texts. As will be shown in Chapter 4, this confusion has far reaching consequences affecting communication as cognitive (adaptive) activity.

Another authoritative source (Trask 1993) defines grammar as, on the one hand, the system by which the words of a language are organized into larger units, particularly into sentences perceived as existing independently of any attempt at describing it, and on the other hand, as a particular description of such a system, as embodied in a set of rules. Correspondingly, the branch of linguistics dealing with the construction of such descriptions and with the investigation of their properties, is also called 'grammar'. Note that 'independent existence' of grammar in the first sense is nothing but a tribute to another axiom of orthodox linguistics, that language is something 'out there', in the real world, waiting to be 'acquired'. However, as Didier Bottineau (in press) observes, "language is not an object, it cannot be acquired." As it were, sentences do not 'exist independently' of languaging humans. Moreover, as descriptive construals, they are cultural constructs, artifacts used in second order activity (or first-order linguistic activity) when we do what we talk about when we talk about language — for example, when we 'analyze' written texts. As it were, we never speak in sentences (typically defined as grammatical, i. e. syntactically organized, units expressing independent and/or complete thought) in normal, routine everyday discourse, and those who try to do so for whatever reason are usually looked upon as weirdoes or freaks. The concept of sentence itself is a second-order cultural construal pertaining to strings of written signs that stand in for words of natural spoken language, and to claim that people speak in sentences is akin to insisting that bees live in hives or wolves run in packs — forgetting that these are linguistic descriptions which we make of something as observers and which have nothing to do with what bees or wolves really 'think' they do. It is not therefore surprising that judgments about 'grammaticality' (well-formedness) of sentences often vary among native speakers of a particular language, because, as Esa Itkonen (2003: 114) observes, "the correctness of sentences is a social fact".

For Steven Pinker (1999: 4) grammar is "a code or protocol or a set of rules that specifies how words may be arranged into meaningful combinations". This code (also known as 'language faculty') sits in everyone's head, and the mind 'works by words and rules' or, more generally, 'by lookup and computation' (*ibid.*, p. 21). A similar view on grammar is held by many linguists; for example, Peter Culicover (2005: 241) writes: "It is a description to a certain level of abstractness of the behavior of a complex dynamical system that is in the head whose function is to compute correspondences between form and meaning". Such views are characteristic of formal approaches to grammar, and the computational metaphor lavishly used in describing language and cognition (and at which we shall yet take a closer look in due time) is the hallmark of generative grammar and first generation cognitive science. On the nativist view characteristic of generativism, as Goldberg, Casenhiser and Sethuraman (2004: 290) observe,

> learning a grammar can be likened to customizing a software package: everything is there, and the learner simply selects the parameters that are appropriate for his environment.

Generative linguists insist on disembodiedness of grammar as a core methodological tenet. This tenet is based on the premise, endemic in modern sciences, that insight into the nature of a totality can be gained by decomposing it into its component parts and studying them separately. For example, a generativist could draw parallels between grammar and, let us say, chemistry, assuming that analysis into components can lead to revealing truths about their totality — especially since chemists do seem to find some utility in divorcing the components of a compound of interest from one another. However, the undeniable utility of such analysis in chemistry does not always help answer the important question about the nature of the unique properties of the compound itself as a whole. The fact that a molecule of water consists of two atoms of hydrogen and one atom of oxygen (H_2O), each of which is a gas, and the fact that hydrogen is a highly flammable substance while oxygen is a component that makes combustion possible, do not of themselves help understand how a particular structural combination of the two elements yields a substance (water) so radically different as to be used to put out fires. "Disembodied" ('formal', 'generative') grammar, understood as something that can be singled out as a universal invariant among human beings, as a totality of molecules that can be formed using a limited set of components, is a fiction, a stark delusion rooted in Cartesian dualism.

The view that grammar should be understood as an innate system of rules governing the functioning of language as a means of communication according to generative principles has been criticized for its biological implausibility (Bates and Goodman 1998; Deacon 1997; Elman et al. 1996; Sampson 1997), and a new approach to grammar as cognition has become remarkably popular (Langacker 1987, 1991; Geeraerts 1988; Goldberg 1995; Heine 1997). While the traditional approach to grammar attempts to provide an explicit set of features of meaning allegedly associated with a given form, within the cognitive paradigm these features are viewed as an assembly of concepts we have about the world which are acquired *via* natural language and *with* (the help of) language. The outcome of this acquisition process is the sum total of our linguistic experiences, of our "romancing the language" — not as a system in a structuralist sense, but as a natural environmental domain inseparable from any other non-linguistic aspect of man's existence. Thus grammar is viewed as a system of categorized patterned cognitive experience (Bod 1998).

There is a general agreement among those involved in cognitive explorations of language and linguistics that the Chomskian view of grammar inadequately reflects linguistic reality for the simple reason that a grammar rule as a specific concept is the product of an observer's interpretive activity and its definition is predetermined by the initial set of assumptions used as points-of-reference in construing this concept, i. e., by the currently accepted and used concepts about language including the concept of 'grammar rule'. Interestingly, this is one of the reasons why language is best acquired without any conscious application of rules even if they are known to linguistic science.

It is a well-known fact that word processors with grammar-checking utilities do not provide a comprehensive content-oriented analysis of grammatical structures used in the text. Thus, although alternative use of indexicals may considerably affect sentence / paragraph / text meaning, there does not seem to be a technical solution to the problem of automated content-sensitive monitoring of grammar in natural language processing. Consider the following sample text:

(2) *The Statue of Liberty is standing on Liberty Island in New York Bay. It will have been built in the last century and is being a symbol of freedom to millions of immigrants going to America.*

Run through Corel WordPerfect Grammatik, it appears to be perfectly correct, though none of the verbs used in it are in the grammatically correct form, that is, they don't conform to the 'rules' that need to be used to make the sentence 'well-formed'. Ideally, a grammar checking utility should be able to yield a processed grammatically correct text such as follows:

(3) *The Statue of Liberty stands on Liberty Island in New York Bay. It was built in the last century and has been a symbol of freedom to millions of immigrants coming to America.*

That this does not happen only proves that there is no basic understanding of how grammar works both on the part of computer software developers and linguists on whose language models programmers base their work. Traditionally, a specific grammar rule is formulated deductively by analyzing relations between signs (the conventional domain of syntax). In the course of such formulation it is usually overlooked that all possible normally observed relations between signs depend entirely on the observer's linguistic experience. In its turn, linguistic experience, by definition, implies other kinds of sign relations which constitute the domains of semantics and pragmatics (cf. Kravchenko 1996) and are empirically rooted in the *interactions of humans with the environment.* As argued by Bod (1998: xi ff),

> the productive units of natural language cannot be defined in terms of a minimal set of rules (or constraints or principles), as is usually attempted in linguistic theory, but need to be defined in terms of a redundant set of previously experienced structures with virtually no restriction on their size and complexity.

As a cognitive mechanism, grammar should not be understood as a set of rules for using signs but as some general principles underlying the mechanisms for reflecting human interactional experience in semiosis; the nature of grammar cannot be explicated without addressing the issue of human cognitive activity (Barlow and Kemmer 2000). With a few exceptions (e.g., Langacker 1987, 1991; Geeraerts 1988; Heine 1997; Dirven and Verspoor 1998), in the past twenty years or so the theoretical principles of cognitive linguistics or cognitive semantics have been applied little if at all to grammatical categories, most of the effort of cognitive linguists being directed to lexical semantics. However, the importance of such application is becoming more and more obvious (Turewicz 2000). It is almost a matter of routine practice today in contemporary analyses of grammatical phenomena to make recourse to such concepts as *observer, experiencer, perceptual domain, personal space*, etc. It is symptomatic of the shift in the theoretical conception of natural language grammar. One such possible conception will be offered in Part Two of this book.

Within the framework of the bio-cognitive philosophy of language (Kravchenko 2003a) the function of language as a system of (written) signs is to accumulate and store categorized experience (knowledge). The meaning of linguistic signs is experiential in nature:

> All conventionalized linguistic expressions (morphemes, words, idioms, phrases, etc.) are connected with "meaning potentials"... A meaning potential is basically a person's memory of the previous uses of a particular expression and can be seen as the union of all the information the person can associate with the expression (Allwood 1999: 1-2).

Meaning emerges from interactions of humans with the environment and others (more on this in Chapter 3). As all our experiences are, ultimately, derived from perception, a scientific study of language is impossible without the study of perceptual processes as meanings are perceptually grounded (Gärdenfors 1999): much of what people say "depends directly or indirectly on their perception of the situation they are in" (Miller and Johnson-Laird 1976: 2). However, the human physiological makeup imposes certain limits on the number and quality of possible percepts, so in all probability grammar is, fundamentally, a relatively simple system for categorizing cognitive experience. Therefore, a theory aiming at explaining grammar (or a part of it) should also be relatively simple and generally comprehensible. Yet prescriptive grammars of particular languages, especially those used in second language acquisition, are notoriously riddled with comprehension problems for foreign language students.

For example, native speakers of Russian have tremendous problems coping with so-called tenses in English. The process of grammar acquisition is from the start hindered by the cognitive hurdle presented by the sheer numbers of English tenses: just because in Russian schools students are taught that there are three tenses in Russian (present, past, and future), they are baffled, sometimes for ever, by the twelve tenses in English, not counting Future-in-the Past and the passive forms (more on this in Chapter 6). Likewise, there isn't a single book on Russian grammar for foreign students, such as native speakers of English, that might in any fathomable way facilitate acquisition of Russian verbal aspect. A similar observation could be made about the English article and word order in Russian sentences. While native speakers of Russian experience tremendous difficulties in 'acquiring' the English article, native speakers of English get lost among the various sentence structures used to express the same propositional content, such as in (4):

(4) 'Ivan find TENSE book A in garden L'

To borrow an example from Kravchenko (2003b: 42), this proposition can have the following *eighteen* syntactic realizations in Russian:

(5)
a. *Ivan nashol knigu v sadu*
b. *Ivan nashol v sadu knigu*
c. *Ivan knigu nashol v sadu*
d. *Ivan knigu v sadu nashol*
e. *Ivan v sadu knigu nashol*
f. *Ivan v sadu nashol knigu*
g. *V sadu Ivan nashol knigu*
h. *V sadu nashol Ivan knigu*
i. *V sadu knigu Ivan nashol*
j. *V sadu Ivan knigu nashol*
k. *V sadu knigu nashol Ivan*
l. *V sadu nashol knigu Ivan*
m. *Knigu Ivan nashol v sadu*
n. *Knigu nashol Ivan v sadu*
o. *Knigu nashol v sadu Ivan*
p. *Knigu v sadu Ivan nashol*
q. *Knigu Ivan v sadu nashol*
r. *Knigu Ivan nashol v sadu*

It cannot be denied that each sentence expresses a meaning slightly different from the meaning expressed in the other seventeen sentences, yet even native speakers of Russian would find this difference difficult to pin down — at least, in some of the sentences. And to ask that of a speaker of English learning to speak Russian by using Russian prescriptive grammars would be to ask the impossible even for the fact that such grammars have nothing to offer by way of explaining examples such as in (5). Yet the mere fact that speakers of one language can learn to speak the other and become fluent in it to an extent when they may even pass for a native speaker, proves that the cognitive mechanisms involved in second language acquisition are not accounted for in prescriptive grammars. Of course, the question one is tempted to ask is, "Why don't grammars do what they are supposed to do, that is, explain grammatical phenomena?" The answer has already been given above: there is no basic understanding in orthodox linguistics of how grammar works. Part Two of this book will try, at least partially, to make up for that.

1.5. Meaning and Communication

Let us now look at the 'meaning axioms' of orthodox linguistics. The claims are that language is a code in which every sign expresses a certain meaning and that the meanings of linguistic signs are given in advance and may be identified and described as abstract invariants on which humans draw in communication as needed by combining them in certain ways to yield meaningful expressions. Chris Sinha (1999) speaks in this connection of the Two Dogmas of Reificatory Semantics, the Dogma of the Autonomy and the Dogma of Compositionality of Linguistic Meaning. As for the function of communication, for the hardened traditionalist it is transfer of meanings. This view is well illustrated by the following excerpt from Pinker (1999: 22):

> Language does express meaning as sound, of course, but not in a single step. Sentences are put together on an assembly line composed of mental modules... One is a storehouse of memorized words, the mental lexicon. Another is a team of rules that combine words and parts of words into bigger words, a component called *morphology*. A third is a team of rules that combine words into phrases and sentences, a component called *syntax*. The three components pass messages about meaning back and forth with the rest of the mind so that the words correspond to what the speaker wants to say. This interface between language and mind is called *semantics*.

The picture drawn by Pinker reminds of the Lego blocks whose simply shaped pieces are designed so as to allow their particular combinations (by the trial-and-error method, of course, or 'passing messages about meaning back and forth') into larger and yet larger pieces of more complicated shapes until — *voilà!* — you've got something you recognize as a car or a house or whatever, which is to say, something *meaningful*. The little Lego pieces themselves do not have any meaning other than to be combined with one another following a certain pattern, while the sample pictures provided with the set of blocks may be viewed as 'meanings' (or 'thoughts') to be expressed with the help of little pieces. In the case of the Lego blocks, both the pieces and the 'meanings' are given in advance, neatly stored in a box until a user has opened it and started to 'express meanings' by putting the pieces together. It is the very picture orthodox linguistics draws of language by insisting that meanings are 'in the mind', in the 'storehouse of the mental lexicon' which consists of words and 'parts of words' (the notorious symbols), and that there is a 'team of rules' to combine these into sentences — and all this happens in the 'mind box'. After meanings (the symbols in the mind) have been 'assembled' as deemed appropriate by the owner of the mind, they are 'expressed as sounds' which carry meanings across to other language users (communicants) and into their mind boxes. This model is known as the 'code model' and is built on the 'conduit metaphor' (Reddy 1979) extensively exploited by orthodox linguistics in discourse and communication studies. On this view, communication is exchange of (coded) information (meanings, thoughts, etc.), that is, *telementation*.

There is, however, one important difference between the Lego game and human language, making the analogy between the two impossible. While in the Lego game both the meanings and the pieces used to express them are *material perceptible things*, in the code model of language meanings are *dissociated* ('disembodied') from acoustic phenomena (physical words), they exist *a priori* as 'mental lexicon' (Aitchison 1987) and become perceptible only through the sound of physical words. On this approach, as Mateusz Oleksy (2002: 59) notes, a communicative event is seen "as a bipolar speech event, composed of two binary relations, one holding between conceptualization and vocalization, and the other going in the opposite direction". Yet it is not, Oleksy argues, as if one speaker encoded an expression and left it in the air and the other speaker picked it up from there and proceeded with decoding it:

> The idea of communicative act as a conjunction of two binary relations resembles the idea that the act of giving someone a present consists in the fact that one person leaves an object somewhere and another one picks it up from there. But the act of giving someone a present which may be realized in a countless number of ways (not necessarily involving a material object, still less a physical transmission) essentially involves the transmission of the right to property. This sort of conceptual analysis led Peirce some time ago to the conclusion that semiotic act, and social act in general, cannot be reduced to any complex of binary (dyadic) relations of whatever type (*ibidem*).

In very much the same spirit, Ross and Spurrett (2005: 643) make the following observation:

> It is to be sure an immensely powerful folk hunch that complex structures are made of little 'things' and that processes decompose into the bringing together of these little things. However, it is not science.

It is not, therefore, surpising that the 'holy grail' of linguistics has not been found, and the problem of meaning has not been resolved. The questions still waiting to be answered are: (1) "What are meanings?", (2) "Where do meanings come from?", and (3) "What is the function of meaning?"

Since Vygotsky (1987), whose ideas became, somewhat belatedly, widely known to Western science only in the last two decades of the 20th century, the view that meaning is some kind of ideal content contained in linguistic signs — which has been the hallmark of orthodox Cartesian linguistics — cannot be accepted as tenable anymore. Vygotsky's thinking followed the idea that the whole (the sum) is greater than the parts, thus transcending both Cartesian and Spinozian thinking. However, his approach has been often misinterpreted, particularly in attempts to resolve the sense/meaning problem by tackling it outside the framework of a "complex synthesis" Vygotsky advocated. For Vygotsky, the point of establishing word meaning as a unit of analysis was to study the *relationship(s) between thought and language* within a perspective conserving the meaning of the whole. Within this framework, the dynamics of the relationship of thought to word meaning may be understood by using functional analysis and by examining the role of word meaning in the process of thinking. In other words, Vygotsky suggested, though perhaps not explicitly enough, what many cognitive scientists today believe to be unquestionably true, that is, that *meanings are not contained in words* (see, e. g., Gauker 2003). The 'whole' of which he spoke surely cannot be understood as the combinatorial result of matching a sound form (the physical word) and an ideal content such as meaning: were it the case, one couldn't possibly speak of any substantial role played by word meaning in the process of thinking. Indeed, the 'container' metaphor from which linguistic semantics is incapable or simply unwilling to break away, makes any productive discussion of the problem of meaning meaningless (no pun intended).

The source of the 'container' metraphor in linguistic semantics can be easily traced to the written language bias and, in more general terms, to the times when linguistics began to establish itself as a science; as the study of language, it began with the analyses (interpretations) of texts constituted by written linguistic signs believed to represent natural spoken language. Indeed, it is easy to imagine how the first naïve linguist was tempted to see graphic patterns on paper (or other material) as 'objects', as something which 'contained' somebody's thoughts he was 'extracting' in the process of reading ('decoding') — forgetting that before he learned to read he had to be taught to recognize particular graphic patterns as signs representing — and quite arbitrarily at that — particular sounds and sound clusters that shape up the physical words of natural spoken language. As pointed out by Nigel Love (2004), while writing involves the deployment of a finite set of graphic units explicitly learned as such, learning to speak does not require the preliminary learning of a finite set of phonic units, which are then strung together to form utterances. Importantly, when we hear someone speak we do not go through the process of matching particular sound sequences (acoustic words) against 'meanings' in order to be able to 'recognize' the speaker's thoughts; nor do we first 'think thoughts' and then look for appropriate sound patterns to express them when we take on the role of the speaker. To believe that this is the case with natural linguistic communication is to believe that there are separate depositories for linguistic forms (words) and meanings somewhere (e. g., in the mind, as adepts of the computational approach to mind would hold). These forms and meanings are matched against one another accoding to specific sets of rules (grammar), which implies that natural language is a kind of code. However, were it the case, there would be no sense in speaking of any problems with understanding that arise in the course of linguistic communication — while we are only far to well aware of how painful the path to real understanding can be. As Love (2004: 536) observes, "successful communication by means of language often seems to involve working round, rather than with, the alleged deliverances of the alleged code". Can we, Love goes on to ask, identify a realm of non-linguistic thoughts or ideas that language encodes?

> Suppose I see in front of me a scene that prompts me to say 'there's a pig in the garden'. Among the multifarious non-linguistic mental happenings or processes that attend my seeing this scene, which is the one that the linguistic utterance represents or expresses or corresponds to? It certainly seems obvious that there are indefinitely many different visual scenes about which 'there's a pig in the garden' would be a sensible thing to say; and conversely, that there are indefinitely many utterances that would be a` propos as remarks to make when confronted with the scene in question. Far from merely encoding some entity that already exists in non-linguistic form, the utterance seems to belong to and to make manifest a cognitive dimension that is *sui generis*. This is borne out by the difficulty we have in even conceiving what it might be to identify a non- or pre-linguistic 'idea' or 'thought', except by representing it verbally. 'There's a pig in the garden' expresses the thought that there's a pig in the garden. But are there two separate items here, the thought that there's a pig in the garden, and the utterance 'there's a pig in the garden'? Of course, one can silently articulate the words to oneself, but a thought is not just a silent articulation (Love 2004: 538).

'The utterance seems to belong to and to make manifest a cognitive dimension that is *sui generis*'. It is this 'cognitive dimension' that Lev Vygotsky posited as a unit of analysis in the *relationship(s) between thought and language.* Natural language acquisition is a holistic process; children can learn to use linguistic constructions appropriately and productively "only by witnessing events in the world and attempting to discern the speakers' communicative intention in referring to these events with a certain pattern of linguistic symbols" (Tomasello 2005: 192). As argued by Gunter Kress (1997), in making meaning with and through signs of various kinds, people — old as well as young — act out of their current interest and put to use in sign-making whatever is to hand. Which is to say that meanings are not pre-given, they emerge in the course of cognitive interaction with the (social) environment. As Christopher Gauker (2003) argues in his critique of the traditional view of linguistic communication, the asymmetry between thought and language fails the use of words as the expression of thought. Thus, the belief that language is a code is a myth (Harris 1981).

In cognitive linguistics as contrasted to formal approaches to language, according to Gilles Fauconnier (1999: 96), "language is in the service of constructing and communicating meaning, and it is for the linguist and cognitive scientist a window into the mind". At the same time,

> [L]anguage is only the tip of a spectacular cognitive iceberg, and when we engage in any language activity, be it mundane or artistically creative, we draw unconsciously on vast cognitive resources, call up innumerable models and frames, set up multiple connections, coordinate large arrays of information, and engage in creative mappings, transfers, and elaborations. This is what language is about and what language is for (*ibidem*).

Not contesting the general cognitivist claim that language is a 'window into the mind', one cannot but feel dissatisfied with Fauconnier's definition of the language function. First, minor terminological differences aside (such as between *thought* and *meaning*), it does not essentially differ from Malinowski's (1927) definition of language as an instrument of thought and of the communication of thought, which itself goes back to Locke (1961). Second, the idea that meanings are 'communicated' brings us back to the 'code model' of language which is empirically untenable because it is nothing but a myth. Characteristically, Fauconnier (1999) admits that our folk theory of language — that is, that meanings are directly contained in words and their combinations, — although extremely useful to us as human organisms in everyday life, is wrong at the scientific level of inquiry, just like other folk theories. That meaning 'is in the language forms' is, for Fauconnier, a general illusion — but then, and it follows from his description of the function of language quoted above, meanings are somewhere else (presumably, in the head), outside language which, in this case, appears to be 'out there'; meanings are pregiven (are they what constitutes the 'vast cognitive resources' we unconsciously 'draw' upon?). This basic contradiction is not removed by positing an 'instrumental' function of language ('language is in the service of constructing meaning'), because it remains unclear how language participates in constructing meanings on which we draw using that very same language. This contradiction could be resolved by turning to Vygotsky's ideas, especially in view of Fauconnier's belief in the necessity of generalizations:

> Within conventional linguistics and philosophy of language, <...> we find a pretheoretical division of components and phenomena, and accepted methods for gathering data which stand in the way of making important connections and generalizations (Fauconnier 1999: 106).

Regrettably, he lets the matter rest there.

Third, and finally, neither Malinowski nor Fauconnier pose the question why people need to construct and communicate meanings — is it something epigenetically superficial, a kind of 'whim' that for some reason hasn't been accounted for, even though it could be done, say, by psychologists, or is it something humans are driven to do by their very biological and social organization as living systems? The questions, thus, are, "What is the (biological) function of meaning?" and "What is the (bio-social) function of communication?" Notably, both in orthodox linguistics and in cognitivism these questions have been hardly addressed at all.

As humans, we become what we are through immersion in the flux of joint activity with others, or *communication*, and in the case of the species *homo sapiens* the uniquely characteristic feature of this activity is *linguistic behavior* or *languaging* outside and without which it is impossible to imagine our very existence. However, there is a strong tradition in philosophy of language to treat linguistic activity as parasitic on the Aristotelian concept of *praxis*. Classical theory views language as something 'out there', a tool, an instrument used to enhance human praxis, overlooking the fact that there can be neither human praxis nor theory outside and without language. In other words, we as humans exist, or *happen,* as Humberto Maturana (1978) puts it, in language and cannot be understood without it.

On the other hand, rather than viewed as a means of exchanging information, linguistic communication should be understood as a process by which speakers *govern* whatever processes go on in the minds of hearers (Õim 1977), as an instrument for exerting influence and coordinating different perspectives (Verhagen 2005). For Herbert Clark (1996), linguistic signs as signals are 'coordination devices in joint activity'; therefore, as Ruth Millikan (2005: 53) argues, "to understand how language works one must look, first, to the cooperative functions that various language forms perform". This view departs from the traditional belief that language is 'denotational', that is, that words 'denote' "persons, things, places, properties, processes and activities external to the language-system" (Lyons 1977: 207). If we view language as the ultimate praxis of humans, we must agree with Mateusz Oleksy (2002) in that it is not based on theoretical knowledge, nor is it determined by rigid, objective norms. Yet it is precisely this rather trivial epistemological premise that orthodox linguistics resists accepting, continuing tenaciously to oppose speech to language, language to thought, linguistic consciousness to non-linguistic, etc. As an example, the speech-language dichotomy as the foundational concept of structuralism is the ultimate manifestation, in a condensed form, of the axiomatic assumptions about language and communication I have already listed: language (*langue*) is out there, we draw on its resources and put them to different uses in speech as needed just like we draw money from a bank account to buy goods, pay bills, or simply waste it away in a casino, somehow forgetting that for the money to be there for us it has to be deposited in the first place.

As far as language ontology is concerned, the situation in contemporary linguistic science is rather curious. On the one hand, the term "natural" as applied to human language is a commonplace label routinely used in language sciences. On the other hand, it is hardly possible to cite an example of a single work where language has been consistently treated as a *natural* (empirical) phenomenon. Understandably, there is a good reason for that, as such an approach would imply taking a holistic stand; that is, it would demand that we look at language as a whole rather than as an assortment of different functional (in the narrow linguistic sense of the term) features treated formally in the domains of syntax, semantics, or pragmatics.

Speaking of interpretation as essentially a holistic enterprise, Hilary Putnam (1983: 149 ff) makes the following observation:

> To interpret a language... involves finding a translation scheme, an 'analytical hypothesis', which is capable of being learned, capable of yielding ready equivalents in our home language to the expressions being translated, and most important, which is such that when we interpret the speakers of the alien language as meaning what the translation scheme says they mean we are able to 'understand' their purposes, beliefs, and behavior. <...> If this is right then the only criteria that we actually have for the 'content' of any signs, or sign-analogs, are our intuitive criteria of successful interpretation; and to *formalize* these would involve formalizing our entire conception of what it is to be human, of what it is to be intelligible in human terms.

In other words, a holistic approach to language is impossible without a coherent philosophical conception of language as a feature of the species *homo sapiens* viewed empirically, that is, as a living system. If one emphasizes the symbolic aspect of language as a system of signs used to 'represent and process knowledge', then the focus of attention shifts to the study of intrasystemic relationships between different types of signs and structures for representing knowledge (whatever those may be), with inevitable controversies about the nature and character of categories, concepts, representations, etc. Meaning in this case is understood as information (cf. Fodor 1998), with linguistic structures as its vehicle. Hence, the theory of meaning developed within this framework is heavily burdened with logical formalizations aimed at providing an explicit well-defined theoretical (as opposed to empirical) model of semantics governed by sets of rules which, according to Noam Chomsky (1965), ultimately constitute the so-called *I-language* as a biologically based feature of the brain (cf.: Jenkins 2000).

If, to the contrary, one emphasizes the interactional features of language as a kind of joint activity which consist of the continuous making of linguistic choices from a wide and unstable range of variable possibilities in a manner which is not rule-governed, but driven by highly flexible principles and strategies, it is only natural to ask, joining Jef Verschueren (1999), how communication in this case can still be possible, particularly, if one subscribes to the claim that the meaning of a linguistic sign in this case is nothing but its use determined by a set of rules. It is, therefore, only appropriate to agree with René Dirven and Marjolijn Verspoor (1998: 14) that "a more comprehensive view of language as a system of signs must also include the human "conceptualizer" and the world as it is experienced by him".

1.6. Summary

The international linguistic community, by and large, is in many ways a traditional community, and although tradition as a socially approved historical form of praxis proven to sustain species viability is not a bad thing in itself, the changing human praxis, especially in the post-industrial era, comes in conflict with tradition on an ever growing scale. Adherence to the orthodox views on language listed, even though partially, in Section 1, is a tribute to the strong tradition in the study of language set by classical philosophy. This philosophy viewed language as something autonomous, something 'out there', a 'thing', a tool of a kind, which humans use to whatever ends just like they use other things or tools, the difference being that language, unlike other tools, is used to do something with immaterial things such as thoughts, ideas and the like. Although over the centuries the tribute to the classic tradition has been paid more than enough, it has not led to any remarkable progress in the study of language, and human praxis has not been noticeably affected by whatever discoveries and theories orthodox linguistic science may boast. Obviously, the classic tradition and its modern heir, analytic philosophy, as the ultimate quest for knowledge about human knowledge, come short of understanding some fundamental things about the world we call 'reality' — again, out of tradition.

As a result, the developmental path of linguistics as a science (should we agree that it is, after all, a science) has been a reiteration, more or less, of analytic 'truths' about the so-called 'objective' world applied to linguistic phenomena (for a more detailed discussion see: Love 2004). As becomes evident from the discussion in this chapter, there is a dire need for language sciences to coordinate efforts in working out a more realistic framework for the study of natural language as a cognitive phenomenon. Given that language, as species-specific joint coordinated behavior of humans, is most appropriately viewed, analyzed and described in biological terms, that is, as a kind of adaptive activity involving interactions in a linguistic domain constituted by the complex dynamics of communicative behavior in real time and space, such framework should integrate various disciplines concerned with the study of different aspects of language, but above all, it should bring together linguistics, semiotics, and biology by placing them on a common epistemological foundation. Some tentative steps in this direction will be taken in Chapter 2.

CHAPTER 2

TOWARD A REALISTIC LANGUAGE SCIENCE

2.1. Introduction

The transition of human society into the new millennium has been marked, among other things, by a growing dissatisfaction felt in the humanities and social sciences. This dissatisfaction has been mostly caused by the slow advances made in the study of the essence of human nature. The steady process of humanization of science observed throughout the 20th century has led to a realization that the quest for 'objective' knowledge of the world is an enterprise altogether unattainable as long as the subject of cognition is not brought into the picture (Chardin 1956; Schrödinger 1959). To quote John Searle (1984: 25), it is "a mistake to suppose that the definition of reality should exclude subjectivity". But for the cognizing human subject to be incorporated into the scientific worldview it needs to be given a status which would be on a par with the status of any other known 'objective' phenomenon. Basically, it means that the human subject should be treated as an empirical phenomenon constitutive of the world this subject conceptualizes and categorizes. Such conceptualization and categorization become manifested (reified) in *language* as human-specific natural (biological, or bio-social) activity. These linguistic manifestations are generally understood as *knowledge structures* or simply *knowledge* which is viewed as the result of cognition as a biologically determined purposeful activity whose goal is to acquire *experience* of the world and to use it to the advantage of humans in their interactions with the environment:

> Knowledge systems <...> unfold from our bio-psycho-socio-linguistic conscious being. Their function is to help us orient (ourselves) in the world and help us act together in the most fruitful way (Brier 2006: 283).

In order for this acquired experience to be used, it must have *meaning* — something that exerts orientational influences on individual human organisms, thus becoming an integral part of the very process of *life*.

The key notions of *language, knowledge, experience, meaning*, and *life* are indispensable in any attempts to penetrate into the essence of humanness — attempts repeatedly made over the centuries by philosophy, biology, psychology, semiotics, linguistics, and other related fields of study. However, impressive as they may seem, current achievements in each of these sciences have not yet provided a set of tools for revealing the true nature of their respective objects of study. The learned community continues to be plagued by the questions, "What is knowledge? What is experience? What is language? What is meaning?" and, ultimately, "What is life?"

It is my firm belief that science will continue to linger in this unhappy state of permanent internal dissatisfaction as long as the term itself continues to be used in a contradictory manner characteristic of modern times; although the word *science*, both in professional and everyday discourse, is used as a generic term (that is, it refers to the systematic knowledge of the world), there is in fact no such thing as *general science*.[2] There is a large array of *particular* sciences (such as physics or chemistry concerned with the study of the physical, that is, 'objective' aspects of the world), which are consistently divorced from the *humanities* and *social sciences* which study the 'subjective' aspects of the world — on the presumptuous ground that the latter do not have a scientifically, i. e., objectively (independently of the human subject) identifiable object of study. It is presumptuous for the simple fact that any knowledge (which is the literal meaning of the word *science*) is the product of cognizing human subjects. This specific product is available to other members of the human community (that is, they become aware of it) mostly in and through language, so any specific branch of knowledge (science) bears the birthmark of language and cannot exist outside of language. In other words, all knowledge is related (Brady 1989/1997). As argued by John Searle (1998: 7),

> because we live in one world, we ought to be able to explain exactly how the different parts of that world relate to each other and how they all hang together in a coherent whole.

Modern sciences have all sprouted from philosophy. The process has taken a long time only to bring scientists to the realization that the more minute the specialization of each separate science, the less overall practical value it offers insofar as the understanding of man, life, and the world goes. This realization heralded the emergence of cognitive science as a new philosophy of life and man. As Marion Brady (1997: 6) observes, "the quality of our lives is largely determined by the quality of the political, economic, social, and religious organizations which structure them". To make any sense of our lives, we must have a good understanding of what it is to be human. From this point of view, we should not so much look for common foundations for the social and cognitive sciences, we should act on the assumption that the two cannot and should not be viewed independently of one another. Central to all sciences must be an understanding that all knowledge is the product of humans as a biological species, therefore, it serves a biological function. And if this function has not been identified, then the purpose of science has not been identified, either.

---

[2] Nor 'general' linguistics, for that matter — despite the fact that General Linguistics has been officially taught at universities as a core linguistic discipline.

Viewed as a physical (material) medium in which most of human communication takes place, language is, undoubtedly, a sign system. To understand, therefore, the language of science we must understand the meaning of signs which, literally, make up science. Signs are the object of study in semiotics, so the intrinsic connection between science and semiotics comes (or should come, to be more precise) to the fore. Yet it has not been the case, and linguistics and semiotics have not been in an intimate relationship over the past century. As a sign system, language is subject to the laws of general semiotics, and the so-called 'linguistic semiotics', concerned with the interpretation of signs in the framework of structuralism, has done nothing more than hinder the development of a comprehensive theory of language as a semiotic system constitutive of human cognitive activity. Since "semiosis encompasses cognition" and

> given that language is a cognitive system and that cognition is a semiotic system, it is odd that two prominent lines of research in contemporary linguistics, one an outgrowth of Peircean notions... and the other a self-described cognitive theory... take little notice of each other (Danaher 1998: 171).

It is not only odd, it is very unhealthy for linguistic science concerned with the problem of human cognition and the role played in it by language.

In this Chapter I am going to argue that the time has, indeed, come for cognitive linguistics and semiotics to embrace each other in reunion like the siblings that have been held apart far too long. I will also argue for the advantages of a biocognitive philosophy of language (Kravchenko 2003a) based on the biology of cognition (autopoiesis) as the theory of living systems (Maturana 1970; Maturana and Varela 1980; 1987). Although the basic theoretical tenets of this framework have not yet gained noticeable recognition in linguistic circles, particularly, in cognitive linguistics as the new paradigm for the study of language, the unquestionably vast explanatory potential of autopoiesis lends it as a methodological stronghold for the effort to find solutions to major linguistic issues that heretofore have evaded more or less consistent and/or comprehensible explanation in the framework of the old traditional paradigm.

2.2. First Generation Cognitive Science

The mainstream cognitive science (Gardner 1985) approach to intelligence is largely computational: intelligent performance is viewed as certain symbolic processes involving representations (Fodor 1975; 1998; Newell 1990; Pylyshyn 1999; Fuchs and Roberts 1999 *inter alia*). These processes account for such cognitive capacities as perception, language acquisition and processing, planning, problem solving, reasoning, learning, and the acquisition, representation, and use of knowledge (Lepore and Pylyshyn 1999). The computational metaphor in today's cognitive science pervades everything, from philosophical psychology to applied tasks in natural language processing. This is only to be expected, since man has been under a spell cast by computer technology for quite some time now.

Although the question, "Does human mind really work like a computer?" has been convincingly answered in the negative (Searle 1984; Lamb 1999), in the predominant cognitive paradigm language is, typically, defined as a sign system for categorizing, storing, retrieving, and processing information. Hence, according to Jerry Fodor (1988: 12), "meaning is information". It is not therefore surprising that the metaphor 'thinking is computation', borrowed by Alan Turing from Hobbes's computational theory of mind (Dascal 2006), is routinely exploited in cognitively oriented linguistic research, forming the basic tenet of contemporary generativism and leading to heated discussions about the nature of mental representations, concepts, categories and, ultimately, meaning (Chomsky 1991; Jackendoff 1996; Deane 1996; Garfield 1997; Croft 1998; Sandra 1998; Cole 1999 *inter alia*).

The basic binary principle that underlies any computational script for intelligent performance is based on the rigid *tertium non datur* premise, that is, an assumption that human reasoning and decision-making is always of the kind "either or" and is governed by the laws of logic. Epistemologically, this approach is the cornerstone of traditional analytic philosophy with its ontological distinction between mind and body (Priest 1991; Schlechtman 1997; Kim 1998). This distinction has borne methodological fruit whose relevance, it appears, has not yet been fully comprehended by the mainstream cognitive science: positing different ontologies for mind and body leads to treating humans as physical entities which serve as vehicles for non-physical (mental) entities. However, these mental entities are housed in the nervous system which is part of the body as a physical entity, and this fact has caused a lot of methodological confusion both among philosophers and linguists.

As a physical object, a human organism is of interest to us mainly because it is a 'container' for the mind which gives it meaning or *value*, but because, presumably, the body and the mind have different ontologies, it remains largely unclear how this value actually comes about. There is obvious parallelism here between analytic treatment of humans and the concept of *sign* as it is defined in semiotics, for both are viewed as binary structures incorporating ontologically different constituents: we speak about the body of a sign and its meaning just as we do about humans, analyzing signs into physical entities (for instance, words of natural language as acoustic phenomena) and mental entities (meanings of words). As far as the theory of meaning, or linguistic semantics, goes, the implication here often is that the body of a linguistic sign is created for the purpose of providing a vehicle for meaning, which leads to an inevitable logical inference that meanings exist before signs. If we subscribe to this assumption, we are faced with the problem of defining meaning as a non-physical entity that exists prior to being assigned to a particular physical entity, such as linguistic sign, through which we become aware of its existence. Thus, we find ourselves in a vicious circle.

The computational metaphor on which almost all modern theories of cognition are built is fraught at least with one grave danger of disregarding the problem of *emerging signification* of signs. Linguistic signs are treated as well-defined binary structures analyzable in terms of form (expression) and content; these structures are routinely used to convey information by making it possible for both the sender (speaker) and receiver (listener) to analyze them, following a certain mutually shared procedure, into quanta of meaning thus extracting semantic values encoded in signs. Such analysis, according to the computational theory of cognition and language, consists in a series of decisions sign users have to make in order to process the encoded meanings correctly (these decisions concern encoding on the part of the speaker, and decoding on the part of the listener). It seems that the entire concept of Artificial Intelligence rests on the premise that, basically, it is possible to develop an algorithmic script that would allow, through application of explicitly formulated logical procedures, to analyze all possible signs into *a priori* defined sets of values contained in them. No matter how appealing it may seem, such an approach fails to take into account the *experiential character* of semiosis as a sign generation process. As Fogh Kirkeby (1997) suggests, we should look at the body-mind as a complex phenomenological dynamical system, including the construction of the environment and the other body-mind systems that make it possible for signification to appear.

Just in the same manner as mind develops (emerges) simultaneously with the development of the body in the course of its functional interactions with the environment, meaning emerges in the course of an organism's functional interactions with physical entities that constitute this environment. Signs in general, and linguistic signs in particular, are epistemologically no different from any other physical entities or phenomena found in an organism's immediate environment. These entities and phenomena may become signs or they may not, and it all depends on whether, in the course of routine encounters with them, an organism assigns *value* to them, which is largely individual rather than social in nature. This value may become socialized to a high degree as a result of humans' interactions with particular kinds of entities or phenomena in a perceptually and experientially shared domain, but it will never lose its individual 'flavor'. It follows from this that neither the set of all possible signs, nor all possible sets of semantic values can be well and exhaustively defined as long as one major factor has not been brought into the picture — *the human experiencer*.

It is highly probable that categorized human experience is patterned in a certain organized way predetermined by human biological makeup. However, it would be wishful thinking to believe that all the diverse lifetime experiences which surely cannot be identical to one another, are categorized following a limited well-defined set of categorization patterns, especially in view of the illuminating insights into categorization processes given by the theory of prototypes. This theory casts serious doubts on the feasibility of binarism as a fundamental concept in understanding intelligence and how mind works, for this concept precludes the very possibility of considering another option for conceptualizing decision-making as an experience-governed indeterminacy reduction procedure based on an assumption that rather than being "either or", it may be "both A and B, with C as a possibility".

A more comprehensive view of language as a system of signs must include the human "conceptualizer" and the world as it is experienced by him, that is to say, language as an object of study should be approached and described *holistically* (Zlatev 1997). As soon as such view is accepted, information becomes the product of cognition as a biological function of a living organism (Maturana 1978), the function of interacting with, and adapting to, the environment.

2.3. Second Generation Cognitive Science

One of the earliest examples of holistic approach to language, which had a noticeable influence on traditional philosophy, was Augustinian conception of language known as the 'Augustinian picture': it included language, mind, and the world (Philipse 1992). A modern philosophical basis for such approach was laid out by John Dewey (1925), who believed that knowledge, mind, and meaning are part of the same world that they have to do with, and that they are to be studied in the same empirical spirit that animates natural science. The idea was further developed by Gilbert Ryle (1949) in his seminal *The Concept of Mind*, where he suggested a materialist approach to the problem as a tangible solution. According to Ryle, anything that exists is material, and the mind is nothing but a totality of bodily dispositions for certain types of behavior. In other words, the mind is described as a complex property of the body and it should be studied as such. Thus, one could say that Ryle had anticipated the emergence of a new paradigm in the study of mind, cognition, and language when cognitive capacities are viewed as *empirical objects of study*.

A holistic approach to language assumes its biological nature: language is viewed as a biological phenomenon uniquely characteristic of the species *homo sapiens* (Bickerton 1990) — not in the generativist sense that there is some kind of *language organ* in the brain which grows with the growth of the organism, but rather in the sense that language as a specific kind of cognitive interactions is an existential/functional domain of humans as biological entities. Attempts of contemporary 'second generation' cognitive science (Lakoff and Johnson 1999), that is, cognitive science based on the fundamental assumption that language is a biological phenomenon in the latter sense, to integrate linguistics and biology in working out a realistic philosophy of language (Devitt and Sterelny 1987), alongside with an experience-based theory of language (Bod 1998) and a usage-based theory of language acquisition (Barlow and Kemmer 2000; Tomasello 2000) are indicative of the new epistemological turn in linguistic science which directly bears on the biological theory of cognition (Maturana 1970) and Peirce's semiotics.

The main assumption of second generation cognitive science is that cognition is characterized by *embodiment* (Johnson 1987; Varela, Thompson and Rosch. 1991), hence the name given to such science — *embodied scientific realism:*

> At the heart of embodied realism is our physical engagement with an environment in an ongoing series of interactions. There is a level of physical interaction in the world at which we have evolved to function very successfully, and an important part of our conceptual system is attuned to such functioning. The existence of such "basic-level concepts — characterized in terms of gestalt perception, mental imagery, and motor interaction — is one of the central discoveries of embodied cognitive science (Lakoff and Johnson 1999: 90).

Although Lakoff and Johnson emphasize that embodied realism "relies on the fact that we are coupled to the world through our embodied interactions" (*ibid.*, p. 93), the focus of attention in embodied cognitive science is on 'conceptual system' and 'conceptual structure', while concepts and reason strongly depend upon the body as "conceptual structure arises from our sensorimotor experience and the neural structures that give rise to it" (Lakoff and Johnson 1999: 77). Accordingly, "second-generation cognitive science is in every respect a cognitive science of the embodied mind" (*ibid.*, p. 78). Consequently, the main effort should focus on the study of neuronal mechanisms in the brain as a key to understanding the biological nature of language and reason. Further progress in the study of language and reason is possible only if linguistic theory is integrated with neural theory, which is to say that man should become the focus of attention, first and foremost, as a *living system*.

Lakoff and Johnson's concept of embodiment seems to be somewhat narrower than originally suggested by Varela, Thompson and Rosch (1991) who attempted to shape up a more comprehensive framework for understanding the nature of language and reason — *enactive cognitive science:*

> We propose as a name enactive to emphasize the growing conviction that cognition is not the representation of a pregiven world by a pregiven mind but is rather

> the enactment of a world and a mind on the basis of a history of the variety of actions that a being in the world performs (Varela, Thompson and Rosch 1991: 9).

The enactive approach assumes that (1) perception consists in perceptually guided action, and (2) cognitive structures emerge from the recurrent sensorimotor patterns that enable action to be perceptually guided:

> The overall concern <...> is not to determine how some perceiver-independent world is to be recovered; it is, rather, to determine the common principles or lawful linkages between sensory and motor systems that explain how action can be perceptually guided in a perceiver-dependent world (*ibid.*, p. 173).

On their view, the body needs to be understood as both physical/biological and, at the same time, as experiential/phenomenological. They assume that both biological interactions and historical interactions influence the mind. Cognition is thus understood as a process of organizing and reorganizing one's own subjective world experience, involving the simultaneous revision, reorganization, and reinterprentation of past, present, and projected actions and conceptions (Davis and Sumara 1997). Modeling language in the light of the enactive paradigm, stresses Didier Bottineau (in press), entails some crucial methodological choices, such as eliminating

> all theoretical prejudice regarding the general categories: the traditional objects of linguistics (*langage, langues, parole, grammar...*) cannot be taken as scientific objects of scrutiny so long as they have not been redefined in terms of dynamic sensorimotor experience.

The concept of cognition as 'a process of organizing and reorganizing one's own subjective world experience' is taken further by Andy Clark and David Chalmers (1998) in their extended mind theory. Using David Kirsh and Paul Maglio's (1994) notion of epistemic action as a kind of action which alters the world so as to aid and augment cognitive processes such as recognition and search, they suggest that epistemic action demands spread of *epistemic credit*:

> If, as we confront some task, a part of the world functions as a process which, *were it done in the head*, we would have no hesitation in recognizing as part of the cognitive process, then that part of the world *is* (so we claim) part of the cognitive process. <...> In these cases the human organism is linked with an external entity in a two-way interaction, creating a *coupled system* that can be seen as a cognitive system in its own right. All the components in the system play an active causal role, and they jointly govern behavior in the same sort of way that cognition usually does (Clark and Chalmers 1998: 12).

The idea of a cognitive system as a coupled system in which all the components play an active causal role, is at the core of the biological theory of cognition, which will be discussed in more detail shortly. What is important, however, is that in second generation cognitive science cognition is viewed as a specifying feature of living systems, and the linguistic community has come close to realizing the necessity of *integration* among the diverse areas of linguistic research which are intrinsically inseparable due to the fact that they all share a common object of study. The scientific exploration of language should embrace all the different aspects by which it is characterized — functional-discursive, psychological, social-cultural, semiotic, etc. At the same time, once we admit that language is an empirical phenomenon and use a holistic non-Cartesian approach in its study, the issue is raised of the physical, namely, biological nature of language. This inevitably invokes interdisciplinarity in linguistic research when the entire complex of the humanistic sciences comes into play.

The issue of interdisciplinarity in science is not an idle one (Heintz and Origgi 2004). Its importance is determined by the fact that, unlike our ancient predecessors, we have achieved a truly spectacular feat of fragmenting our knowledge of the world and the place of man in it to the degree when it is difficult, well nigh impossible, to find common ground for areas of knowledge accumulated by various particular sciences, even when they are 'united' by the common object of study (cognitive linguistics and semiotics being a ready example). Moreover, philosophy itself, usually defined as the science of the general laws of nature, society, and reason, is intrinsically flawed with contemplativeness for it has not as yet borne the fruit of universally accepted truths about what has been known so far. Specifically, there may be no unanimously agreed perspectives on what knowledge is as long as the problem of meaning (in particular, linguistic meaning) remains unresolved.

This thought may be described as one of the main themes of Michael Devitt and Kim Sterelny's (1987/1999) 'realistic philosophy of language' which is attainable only when and if people are seen as part of the natural world since humans are part of the animal world, part of the biological world:

> Any linguistic facts there are must be, ultimately, physical. Semantic notions like meaning, truth, and reference can be used only if they can be explained in nonlinguistic terms; they are not primitive... We might, for example, hope to explain them in psychological terms; then, hope to explain the psychological in neuroanatomical and biochemical terms; then, explain those in physical and chemical terms (*ibid.*, p. 10ff)

Thus, conceptually, Lakoff and Johnson's *Philosophy in the Flesh* is a further development of the 'biological' theme in contemporary philosophy of language. The issue has also been recently tackled by Talmy Givón (2002) in his *Bio-Linguistics,* though some of the questions about the relation between cognition, adaptation and neural structure have been raised earlier (Givón 1995). The appearance of *Bio-Linguistics* is symbolic of the beginning integrative processes already mentioned. Of course, it must be noted that *bio-linguistics* as conceived by Givón differs conceptually from biolinguistics as it is understood by Chomsky and his ideological followers. This difference relies on the kind of answers one gets to such questions as, for example, (1) What constitutes knowledge of language?, (2) How does knowledge of language develop in the individual?, (3) How did knowledge of language evolve in the species?

In generative linguistics as the mainframe of the first generation cognitive science, central to answers to these questions is the notion of *linguistic competence* — an abstract idealization of a native speaker's knowledge of the grammar of his language. However, it does not seem to be a matter of dispute anymore in contemporary science that the notion of competence may be used as an analytic tool only if a strict and non-arbitrary divide is possible between what is often referred to as *linguistic* and *non-linguistic* knowledge. Yet "there is no hard and fast dividing line between linguistic and non-linguistic (encyclopaedic) knowledge, since such division fails to recognize the fact that language is part of our world" (Paradis 2003). That is why, argues Givón (2002: xvi), Chomsky's idealized *competence* is empirically utterly untenable. For Givón, bio-linguistics is a study of language based on an understanding that, being a uniquely human feature, it is intrinsically biological, and in the study of biological organisms "the relevant perspective for studying and classifying structures is their respective adaptive functions" (*ibid.*, p. 205).

So, the obvious tendency of contemporary linguistics to integrate with biological science signals the emergence of a new paradigm in cognitive science of which language is an integral part, namely, the *biological paradigm.* In this paradigm, language is seen as a natural biological phenomenon bearing on the adaptive function of humans as living organisms. This view marked the International Workshop *The Linguistics of Biology and the Biology of Linguistics* which took place in Mexico in 1998 — suffice it to mention the titles of some presentations: *Grammars and gene regulation* (J. Collado-Vides), *The role of self-organization in language* (D. Demolin), *Biology and language* (M. Gopnik), *The grammatical formalization of atomic DNA sequences* (R. Hofestadt), etc. So one can characterize the new direction taken by non-Cartesian cognitive science as *biocognitive,* i. e. we should speak of the *biocognitive philosophy of language* (Kravchenko 2002a).

2.4. Biology of Cognition and Language

2.4.1. An Alternative to the Computational Model of Cognition

However, before we begin to do so we must clearly define our basic assumptions about cognition as a *biological phenomenon;* hence the necessity to recount the basic tenets of the biological theory of cognition or *autopoiesis* (Maturana 1970) as an alternative theoretical framework "both intriguing and subversive" as "it challenges orthodox cognitive science" (Boden 2000: 142, 143), competing with the computational model of cognition and language. The basic tenets of this theory are summed up as follow:

(1) Cognition is a biological phenomenon and it can be understood only as such; any understanding of the cognitive domain from the point of view of epistemology involves understanding it from the point of view of biology. To attain such understanding, two questions must be posed: *What is cognition as a function?* And *What is cognition as a process?*

(2) Everything said is said by an observer to another observer who can be himself or herself. Hence it follows that *in order to understand cognition as a biological phenomenon, an observer and his role in cognition must be taken into account and explained.*

(3) Living systems are unities of interactions which exist in an environment. From a purely biological point of view they cannot be understood independent of their niche (the part of the environment with which they interact). Likewise, the niche cannot be determined independent of the living system which specifies it. A living system is characterized by *circular organization.* This circular organization produces a homeostatic system whose function is to produce and sustain the circularity of organization; the realization of this function depends on the *components* which determine it and the continuous production of which is sustained by the circular organization. All the particular features of various kinds of organisms overlay and support this fundamental circularity by sustaining its uninterrupted continuity in sequences of interactions with the constantly changing medium. Living systems are inferential systems, and their domain of interactions is a cognitive domain.

(4) An observer observes an organism and the environment at the same time, and he distinguishes as his niche that part of the environment which, according to his *observations,* is a part of his domain of interactions. Therefore, whereas to him the niche *seems* to be part of the environment, to an observed organism the niche constitutes a domain of its interactions in its unity and as such it cannot be a *part* of the environment which belongs exclusively to the cognitive domain of an observer.

(5) A cognitive system is a system whose organization determines the domain of interactions in which it can act meaningfully to sustain itself, and cognition is an actual (inductive) conduct or behavior in this domain. *Living systems are cognitive systems, and life as a process is a cognitive process.* The nervous system enlarges the cognitive domain of a living system making it possible to interact with "pure relations"; it does not create cognition.

(6) The enlargement with the help of the nervous system of the cognitive domain makes possible non-physical interactions of organisms in which interacting organisms orient each other with regard to interactions within each other's cognitive domain. This is the basis of communication.

(7) In so far as an organism's niche is a set of all classes of interactions an organism is capable of, and an observer observes an organism in the environment which he himself specifies, to him any behavior of an organism appears as an actualization of the niche, that is, as a first order description of the environment (Description). The relation between the behavior and the niche belongs solely to the observer's cognitive domain.

(8) One organism can modify another organism's behavior in two basic ways:

(a) by interacting with another organism so that both organisms become interconnected in such a manner that the subsequent behavior of one of them is strictly determined by the preceding behavior of the other; such is the case, for example, in courtship;

(b) by orienting another organism's behavior with respect to a part of his domain of interactions different from the part which includes this interaction, while the orientation of the latter may be compared with the orientation of the orienting organism. This can occur only if the domains of interactions of both organisms coincide to a large degree. In this case, there are no interlocked concatenations of behavior because the subsequent behavior of both organisms is determined by the outcome of independent, albeit parallel interactions.

In the first case one could say that the organisms interact, in the second, that they communicate. The second case is the basis for any linguistic behavior.

(9) In the orientational interaction of the first organism its behavior as a communicative *description* causes specific states of relative neuronal activity in the nervous system of the second organism. This state of activity embodies the relations generated by interactions, and is a representation of the other organism's behavior (Description of his niche) as a connotant of the first organism's orientational activity. This representation may be interpreted by the nervous system as a unity of interactions so that the other organism, if it has a respective capacity, can interact with representations of its own Descriptions of the niche as if they were

independent entities. All this leads to the establishment of one more domain of interactions (and, accordingly, an additional dimension of the cognitive domain), a domain of interactions with representations of behavior (interactions), including orientational interactions as if these representations were independent entities within the niche, the establishment of a *linguistic domain.*

(10) Natural language arises as a new domain of interactions in which modifications of an organism are caused by its own *descriptions* of its interactions embodied in states of activity of the nervous system. Natural language is, of necessity, generative as it is the result of recursive use of one and the same operation (which is a neurophysiological process) to the outcome of such use. New sequences of orientational interactions (new sentences) within a consensual domain are, of necessity, understood by the hearer (they orient him) because every component in these sequences has specific orientational functions as a member of the consensual domain which the hearer helps determine. Hence the following important inference: *language is connotational, not denotational.* Once it is accepted that the function of language is to orient an organism being oriented in its cognitive domain without reference to the orienting organism's cognitive domain, it becomes clear that there is no information transfer through language.

In short, as formulated by Peter Harder (1999: 202) — who, remarkably, doesn't make any reference to autopoiesis, — "cognition is causally grounded in the biology of the organism in which it belongs, and functionally grounded in the interaction between organism (including brain) and environment". As may be seen from the short exposition above, the biological approach to cognition and language as *cognitive processes* which specify living systems differs essentially from the conventional representational paradigm. Whereas for the majority of students of the mind the notion of cognition is tied to the idea of knowledge as processed information stored in mental structures accessible for identification and analysis via language as a representational (denotational) sign system, in the biological theory of cognition it is not a means to acquire knowledge of an objective reality but serves an active organism in its adaptation to its experiential world. Information in this framework is understood as being constructed and codependent rather than instructional and referential (Murphy 1992; cf. Glasersfeld 1995); it is viewed as the capacity of certain physical entities of presenting alternative configurations and, consequently, of exerting different actions in regard to other components or the whole system (Moreno, Merelo and Etxeberria 1992). Thus the tenets of autopoiesis call in question the empirical feasibility of not only cognition and its informative and communicative aspects as they are understood by the mainstream cognitive science, but also of the well established theoretical frameworks in the studies of meaning and representation mentioned in Chapter 1.

2.4.2. Explaining the Biology of Language

Maturana's views on language as specific interactions in the cognitive domain are laid out in his essay *Biology of Language: The Epistemology of Reality* (Maturana 1978), which also contains definitions of the key autopoietic notions such as "observer", "unity", "organization", "structure", "property", "space", "interaction", "autonomy", "adaptation", "perception", etc. The first part ("Epistemology") includes the initial assumptions necessary for the scientific explanation of language. In particular, it is emphasized that

> science is a closed cognitive domain in which all statements are, of necessity, subject dependent, valid only in the domain of interactions in which the standard observer exists and operates. As observers we generally take the observer for granted and, by accepting his universality by implication, ascribe many of the invariant features of our descriptions that depend on the standard observer to a reality that is ontologically objective and independent of us. Yet the power of science rests exactly on its subject dependent nature, which allows us to deal with the operative domain in which we exist. It is only when we want to consider the observer as the object of our scientific inquiry, and we want to understand both what he does when he makes scientific statements and how these statements are operationally effective, that we encounter a problem if we do not recognize the subject dependent nature of science (p. 29).

Therefore, argues Maturana, to give a scientific description of the observer as a system capable of descriptions (language), the subject dependent nature of science must be taken as a starting point. This basic assumption, which is the cornerstone of a new epistemology developed by Maturana, is in sharp contrast with the basic methodological principles that characterize the so-called "objective" science as a legacy of the rational paradigm in philosophy. Because the omnipresent classical mind-body distinction has been so deeply ingrained in philosophical thought, the idea that the so-called external world we perceive is not independent of us, that, on the contrary, it is dependent on us in such a manner that the entities we perceive (such as mountains and oceans and trees and so on) are the end results of our perceptual processes, and as such they are in our phenomenal or mental space (Imoto 2004b), has a tang of heresy to anyone with a formal educational background. However, this assumption, essentially, is nothing more than an epistemological extension of Einstein's general principle of relativity against which modern science seems to have no objection. While an original enterprise on its own, Maturana's approach to science and scientific explanation is, at the same time, reminiscent of the ideas put forth earlier by Charles Peirce, John Dewey, Charles Ogden and Ivor Richards, Charles Morris, Bertrand Russell, Gilbert Ryle, Erwin Schrödinger and many others who believed that there may be no science or scientific explanations of reality without the explanation of man as a living system.

It was natural scientists who began to speak of the necessity to take into account the observer when formulating their scientific outlook on the physical world (Chardin 1956; Schrödinger 1959), and it is in the humanistic sciences, far too long dominated by analytic philosophy, that the subjective nature of science has been haughtily dispensed with as a logically untenable and epistemologically mystifying extravagance, as something *abnormal* for which normal science has no room. So it is no wonder that anything else that the autopoietic theory has to offer instantly becomes inconsequential because of the "absurdity" of the starting premise. In its quest for the *ultima ratio* 'normal' science continues to remain in the grip of *vitalistic* explanations of the observed phenomena in the world, to which Maturana opposes *mechanistic* explanations:

> In a mechanistic explanation the observer explicitly or implicitly distinguishes between a system and its components, treating the system and the components as operationally different kinds of unities that belong to disjoint sets that generate nonintersecting phenomenic domains. [...] The reality described through mechanistic explanations... implies the possibility of an endless generation of nonintersecting phenomenal domains as the result of the recursive constitution (organization) of new classes of unities through the recursive novel combinations of unities already defined. For epistemological reasons... mechanistic explanations are intrinsically nonreductionist. ...
>
> ...Vitalistic explanations... do not distinguish between the phenomenal domain generated by a unity and the phenomenal domain generated by its components. The reality described through vitalistic explanations is, necessarily, a reality of a finite number of phenomenal domains. For epistemological reasons... vitalistic explanations are intrinsically reductionist (Maturana 1978: 30).

An example of vitalistic explanation attempted by first generation cognitive science in regard to cognition will be considered later in this chapter, where it will be shown that just because a living system as a composite unity exists in two nonintersecting phenomenal domains, persistent attempts to identify cognition (and language, in the case of humans) as internal processes taking place in the brains of higher order organisms cannot be but futile. Conversely, mechanistic explanation — as, for instance, exemplified by the distributed approach to language and cognition when emphasis is laid on their complex dynamics (see Section 2.6 in this chapter) — focuses on coupling between brains, bodies and the world (Cowley 2007).

In the section "Living and Nervous Systems", Maturana defines living systems as autonomous entities even though they depend on a medium for their concrete existence. This autonomy is the result of their organization as systems in continuous self-production. Such systems are called *autopoietic systems*. An autopoietic system is a closed dynamic system in which all phenomena are subordinated to its autopoiesis and all its states are states in autopoiesis. This conclusion has fundamental consequences for the understanding of the autonomy of living systems, their ontogenetic and evolutionary adaptations to the medium through selection of the structures that permit the autopoiesis of the living system, as well as for the explanation of the nervous system as a system, behavior as the result of structural coupling between an organism (a dynamic system) and the medium, and the difference between instinct and learning.

After the explanation of the organizational principles of living systems, Maturana goes on to the definition of language as a property of an autopoietic system ("Language and Consensual Domains"). The starting point here is the notion of a *consensual domain* as the domain of interlocked conducts that results from ontogenetic reciprocal structural coupling between structurally plastic organisms:

> ... Language is currently considered to be a denotative system of symbolic communication, composed of words that denote entities regardless of the domain in which these entities may exist. Denotation, however, is not a primitive operation. It requires agreement consensus for the specification of the denotant and the denoted. If denotation, therefore, is not a primitive operation, it cannot be a primitive linguistic operation, either. Language must arise as a result of something else that does not require denotation for its establishment, but that gives rise to language with all its implications as a trivial necessary result. This fundamental process is ontogenetic structural coupling, which results in the establishment of a consensual domain.
>
> Within a consensual domain the various components of a consensual interaction do not operate as denotants; at most, an observer could say that they connote the states of the participants as they trigger each other in interlocked sequences of changes of state. Denotation arises only in a metadomain as an a posteriori commentary made by the observer about the consequences of operation of the interacting systems. If the primary operation for the establishment of a linguistic domain is ontogenic structural coupling, then the primary conditions for the development of language are, in principle, common to all autopoietic systems to the extent that they are structurally plastic and can undergo recursive interactions.
>
> Linguistic behavior is behavior in a consensual domain. When linguistic behavior takes place recursively, in a second-order consensual domain, in such a manner that the components of the consensual behavior are recursively combined in the generation of new components of the consensual domain, a language is established (p. 50).

A consensual domain is the domain of interactions as interlocked conducts resulting from ontogenetic reciprocal structural coupling between structurally plastic organisms.[3] In more familiar terms, a consensual domain may be described as an existentially shared physical context of interactions. The concept of 'consensual domain' is important in understanding the biology of cognition and the cognitive nature of linguistic behavior not as solely centered on individual organismic organization, but grounded in interactional behavioral patterns (interactions with other observers). Thus, the key concept of 'consensual domain' is very close to the *ecological perspective* on 'talking together' as outlined by Hodges (2007), who cites Verbrugge (1985) as the first to offer an ecological account of language by suggesting that "the best observation point" for viewing language is to see it as "a form of social interaction". In a special issue of *Language Sciences* dedicated to the cognitive dynamics of distributed language, Ross (2007) observes — very much in the spirit of autopoiesis — that similar public linguistic representations cue similar behavioral responses in individuals with similar learning histories, as a result of conventional associations established by those similar histories.

An understanding of language as behavior in the consensual domain gives a new perspective on natural language grammar (syntax) and eliminates the much discussed problem of ambiguity, whether structural or semantic. Considering that the behavior of organisms in a consensual domain is a set of recursive operations, the surface structure or grammar of a given natural language may be only a description of regularities in the structural coupling between the elements of consensual behavior. As a matter of fact, says Maturana (1978), this surface syntax may be whatever since its development would be a matter of historical contingency, not of structural necessity determined by a particular physiology of an organism. To an observer linguistic interactions appear as semantic and contextual interactions. Yet what takes place in the interactions within a consensual domain is strictly structure-determined, interlocked concatenations of behavior. Therefore,

> for an observer who does not know the structural states of the linguistically interacting organisms, the outcome of a particular linguistic interaction may seem ambiguous, as if the actual syntactic value of a particular linguistic conduct were determined by some internal, not apparent, rule. Yet for each of the actual linguistically interacting organisms there is no such ambiguity. Their internal structure... is at any moment determined by their previous interactions and by their previous independent structural dynamics in a non ambiguous manner. Therefore, the context on which the outcome of a linguistic interaction depends is completely determined in the structure of the interacting organisms, even if this is unknown to the observer. The overheard sentence, "They are flying planes," is ambiguous only for

---

[3] The notions of 'ontogenetic structural coupling' and 'consensual domain' are conceptually consonant with Clark and Chalmers' (1998) and Cowley's (2004a) distrubutionist account of cognition and Cowley's (2002) integrationist account of language, both of which defy orthodox myths about language.

the observer who wants to predict the outcome of the interaction with insufficient knowledge of the structural state of the speaking organism (p. 53).

This point is well illustrated by how orthodox linguists, especially those whose mother tongue is English, analyze "structural ambiguity" (Altmann 1998) in regard to certain types of sentences. To borrow an example from my own earlier work (Kravchenko 2003a, pp. 122 – 124), in a popular college textbook on linguistics (Fromkin and Rodman 1998), which to that date had survived six editions since 1974, students were introduced to the concept of ambiguity with the help of a comic strip about the Indian Chief and the Trader. The Chief sees the sign on a tepee:

BOUTIQUE
Ask for the synthetic buffalo hides

and approaches the Trader with the words:

I'd like to see the synthetic buffalo hides

The Trader scribbles an answer on a piece of paper which reads:

Yes, sir, should I stumble on a herd of
synthetic buffalo you'll be among the first to be notified

The humor of the joke lies in the different interpretations of the meaning of the expression *synthetic buffalo hides*:

(1) 'synthetic hides of buffalo' (Chief's interpretation),
(2) 'hides of synthetic buffalo' (Trader's interpretation).

A linguistics student learns that this is possible because of the characteristic feature of the English syntax which relies heavily on positional (structural) relations between parts of the sentence. A linguist undertaking an analysis of such a phrase from the structuralist viewpoint sees two possibilities for its semantic structure: it can be either *synthetic [buffalo hides]*, or *[synthetic buffalo] hides*. However, in doing this such a linguist ignores the obvious — that in a normal situation in real life it would not occur to anyone to interpret this phrase in the second sense for the simple reason that in the world as we know it there are no synthetic buffalo whose hides might be the object of trade, while something made of synthetic fiber that looks like a buffalo (or some other animal) hide is a very common thing. Remarkably, the author of the comic strip reveals a very keen sense of language and to him the difference between the spoken and written language (and their analyses, respectively) is obvious, otherwise he would not have made the Trader give a *written* answer to a *spoken* question.

Or take another classical example of 'structural ambiguity':

(6) *Flying planes can be dangerous.*

It is only in the unclouded mind of the structuralist, to whom language is a self-contained autonomous system, that the sense 'planes which fly' may be a possible interpretation of the expression *flying planes* in the above sentence. To an average language user, who knows anything at all about planes and very little about linguistics, such an interpretation would be at least weird if not ridiculous, because his everyday life experience tells him that the chances of a plane high in the sky falling on his head are, for all practical purposes, nil. At the same time, the very same experience tells a pilot, for example, that being in the cockpit of an airborne plane means being in a potentially dangerous situation because, unfortunately, airplane crashes happen and people die in them. The point here is that to a language user, who routinely engages in acts of natural linguistic communication, elimination of "ambiguity" as a problem does not arise at all, because in his interpretation of the observed linguistic behavior of others he relies on his so-called 'extralinguistic' knowledge about whether a particular type of situation is possible. As observed by Akmajian et al. (1990: 312), "in general we do a good job of picking the appropriate reading of an ambiguous expression. To overcome ambiguity, the hearer presumes the speaker's remarks to be *contextually appropriate*".

Taking the consensual domain of interactions of organisms as fundamental in the construal of a new epistemology allows to make the next step and define reality (in the section titled "Reality") as a "domain specified by the operations of the observer." Finally, in "Conclusion", we read:

> ...All that remains is the observer. Yet the observer does not exist alone, because his existence necessarily entails at least an other being as a necessary condition for the establishment of the consensual domain in which he exists as an observer. However, what is unique to each observer and makes each observer stand alone, is, on the one hand, his or her experiences, which remain necessarily secluded in his or her operational closure, and, on the other hand, the observer's ability through second-order consensuality to operate as external to the situation in which he or she is, and thus be observer of his or her circumstance as an observer (pp. 60-61).

This 'second-order consensuality' is a constitutive feature of the domain of language as interactions with descriptions (patterns in behavior) of representations (relative states of neuronal activity) which, being semiotic phenomena, are rooted in our experience of the world. The mechanism whereby it becomes possible for language to function as an extension of the human sensorium, will be described in Chapter 3.

As with any idea that revolutionizes a particular area of knowledge, and exactly because it serves this particular purpose — to set a new direction in thinking about things which appear to be well-known because they have been the object of scientific inquiry for a substantial period of time — autopoiesis may not offer a complete picture of cognition and language in the minutest detail. After all, Humberto Maturana is, originally, a biologist, and he has been explaining cognitive phenomena from a biological basis. This is, probably, why, while speaking about the biology of language, he seems to leave open the question of meaning-making which is the core issue in cognitive science. As stressed by Ezequiel Di Paolo (2005), although autopoiesis provides a systemic language for speaking about intrinsic teleology, its original formulation needs to be elaborated further in order to explain meaning-making. This has been attempted by Weber and Varela (2002) who tried to link autopoiesis and meaning-making through the instauration of a natural perspective from which encounters with the world are intrinsically meaningful for the organism. While describing this effort as a 'promising starting point', Di Paolo points out some of its shortcomings, particularly Weber and Varela's "unspoken reliance on apparent implications of autopoiesis which are not elaborated in the theory and do not immediately follow from the original formulation" (Di Paolo 2005: 433). By examining the primary formulation of autopoietic theory and the distinct concepts of robustness and adaptivity, Di Paolo concludes that autopoiesis implies only the former while sense-making needs the latter. However, even though it may appear that structural coupling is overplayed in the autopoietic tradition because human cognition is not just biological but biocultural, the very idea of structural coupling cannot be divorced from the concept of consensual domain of interactions (which is to say, culture) in which this coupling occurs. The role of culture is not denied by autopoiesis, it is implied, even though — Di Paolo is right here — not elaborated. Cultural aspects of meaning-making are explicitly addressed by researchers in the biology and dynamics of cognition (Tomasello 1999; Zlatev 2003; Tomasello and Rakoczy 2003; Cowley 2004a, 2004b; Cowley, Moodley and Fiori-Cowley 2004 *inter alia*).

2.4.3. Methodological Implications

The advantage of the biological approach to cognition and language over the orthodox Cartesian framework is that it resolves the 'vicious circle' problem in explanation by proposing an explanation of an organism as a living system with a *circular organization*. Circularity is viewed as a specifying property of living systems which are described as *autopoietic systems*. Accordingly, language may be defined as a domain of cognitive communicative activity (signaling behavior) in the course of which linguistic and 'non-linguistic' signs are generated to sustain this activity. It is, therefore, not surprising that circular organization as a specifying feature of a living system finds its manifestation in language. It becomes strikingly obvious in semantics as the study of meaning of linguistic signs: all attempts to provide a hierarchically organized linear system of relationships among signs that would explicitly and sufficiently demarcate their meanings, have invariably turned futile, incapable of resolving the so-called Frege's puzzle (Salmon 1986; Taschek 1992). The distinction between the *sense* and the *meaning* of a sign (which, in turn, stems from the distinction between language as a sign system and language as an activity), though useful for specific analytic purposes, cannot help understand the way language *works*.

The crucial difference between the traditional and autopoietic views of language is that the latter assumes its connotational, rather than denotational, nature. Correspondingly, the key notion of *representation* proposed by Maturana is also different from the traditional one. Representations are relative neuronal activities characterizing the state of an organism's nervous system as a structure-determined system; because of this, the sequence of changing relations of relative neuronal activity (description) that appears to the observer as determining a given behavior, is not determined by any functional or semantic value (or the so-called 'conceptual structure' understood to be composed of hierarchically arranged abstract meaning primitives – see Wilkins 2005) that the observer may ascribe to such a behavior but is necessarily determined by the structure of the nervous system at the moment at which the behavior is enacted. It follows that adequate behavior is necessarily only the result of structural matching between an organism (dynamic system) and its medium (which comprises other organisms structurally matched to *their* medium). This conclusion, by showing the futility of decades-long philosophical discussion about the nature of dualistic entities called 'mental representations' gives the entire philosophical discussion about the nature of mental representations a genuinely scientific (naturalist) angle and is a giant step toward understanding consciousness and cognitive (mental) processes (more on this in Chapter 3).

The notion of consensual domain in which *languaging* takes place is a pivotal point in understanding the nature of language as it allows to see its basic biological function: languaging is aimed at monitoring an organism's environment by modifying other organisms' behavior. Since representation, meaning, and description are notions that apply only and exclusively to the operation of living systems in a consensual domain, the entire problem of meaning takes on a new perspective, calling for a revised dialectics of knowledge. Meaning in autopoiesis is not something that exists 'out there' waiting to be discovered, identified, and 'harvested' — an undertaking that different semantic theories developed within traditional analytic philosophy have notably failed to do. Instead, as emphasized by Francisco Varela (1992:14),

> living beings and their worlds of meaning stand in relation to each other through *mutual specification* or *co-determination*. Thus what we describe as significant environmental regularities are not external features that have been internalized as the dominant representationalist tradition in cognitive science... assumes. Environmental regularities are the result of a conjoint history, a congruence which unfolds from a long history of co-determination.

The core tenet of autopoiesis, *Everything said is said by an observer to another observer who can be himself or herself,* implies that every utterance, in virtue of being a sign, requires an interpretation based on the interpreter's (= observer's) experience in general, and experience of linguistic signs in particular. Since languaging is a circular interpretative process in the course of which a human organism tries to maximize the effects of its enaction of the environmental niche it occupies in order to better adapt to the environment, any empirical entity capable of producing enactive effects is viewed as a *meaningful* entity, and the circle of interpretation includes both linguistic and non-linguistic entities (signs) whose ontologies are different, but which epistemologically stand to one another in a relation of *reciprocal causality*. This means that in the course of knowledge acquisition as a life process, cognitive processing (interpretation) of linguistic signs as elements of a communicative description of representations of enactions depends on the interpretation of enactions themselves.

However, any representation presupposes categorization in the sense that, being a closed structurally determined autonomous system characterized by continuous self-production, the nervous system is a network of productions (and disintegrations) of components that: (i) recursively participate through their interactions in the realization of the network of productions (and disintegrations) of components that produce them; and (ii) by realizing their boundaries, constitute this network of productions (and disintegrations) of components as a unity in the space they specify and in which they exist (Maturana 1978). Put in plain words, relative neuronal activity is constitutive of categorization (individuation) processes which are largely determined by interpretation as a kind of interaction with communicative (orientational) representations. Therefore, in the case of languaging as a cognitive domain, it is appropriate to speak *of interpretation of interpretation*.

The notion of consensual domain is also a very powerful tool in working out an empirically sound theory of semiotics whose foundations were laid out by Charles Peirce (1958). Peirce's famous trinity of icon, index, and symbol has been somewhat misinterpreted in semiotic literature as a hierarchical system, which Peirce never claimed it to be: rather, they stand to one another in a relationship of degree to which each of them possesses *indexical properties.* Since indexicality is, to use autopoietic terminology, a consensual property by definition, it only stands to reason that, to quote Umberto Eco (1984: 1),

> the concept of sign must be disentangled from its trivial identification with the idea of coded equivalence and identity; the semiosic process of interpretation is present at the very core of the concept of sign.

To put it bluntly: meanings (that which exerts orientational influences on humans in their cognitive domains of interactions) *are not contained in signs* (more on this in the next chapter).

If we focus on meaning as a cognitive phenomenon, the meaning of sign cannot be defined other than a certain associative potential which is basically a person's memory of the previous uses of a particular sign (Allwood and Gärdenfors 1999), that is, the meaning of a sign is specified and co-determined in the course of interactions in a consensual domain. As Peter Harder (1999: 294) puts it,

> meaning stands on the shoulders of the whole motor, perceptual and conceptual powers of the individual and draws on all this to create something on a higher level of complexity that includes embeddedness in trans-individual processes.

In other words, an entity becomes a sign by acquiring *value* which emerges as the result of such cognitive interactions. Consequently, as was pointed out in Chapter 1, just as a word, which itself is a physical entity, can be a sign of another entity, *any physical entity can be a sign of a word.* Circularity and reciprocal causation as specifying properties of a living human organism result in *the semiotic multiplication of the world.* The reality of these multiple worlds is something that modern theories of meaning should take into account.

2.5. Semiotics and Cognition

2.5.1. Semiotics as Metascience

Exploration of the nature of language as an empirical phenomenon related to the functioning of living organisms as biological systems appears to be a growing tendency in modern linguistics. Today, the majority of research into the problem of the origins of language is done on the assumption of the evolutionary theory of its genesis and development (Bickerton 1981, 1990; Pinker 1995a; Hurford et al. 1998; Loritz 1999; Jackendoff 1999 *inter alia*). Such approach implies that language ability is a biological feature of a living (human) organism which arises and develops in the course of an organism's adaptive activity and hence possesses an adaptive function. One of the first to point this out was Charles Morris (1945: 32):

> The response to things through the intermediacy of signs is... biologically a continuation of the same process in which the distance senses have taken precedence over the contact senses in the control of conduct in higher animal forms; such animals through sight, hearing, and smell are already responding to distant parts of the environment through certain properties of objects functioning as signs of other properties. This process of taking account of a constantly more remote environment is simply continued in the complex processes of semiosis made possible by language, the object taken account of no longer needing to be perceptually present.

Morris' novel idea of the biological nature of semiosis began to take ground in modern science only a few decades later, when the study of animal behavior (ethology) laid a foundation for what later became known first as *zoosemiotics* and then as *biosemiotics.* The foundations of general semiotics laid by Charles Peirce and further developed by Charles Morris, after an inexplicably long period of general neglect on the part of the 20th century linguists, are again coming to the fore in contemporary research. This has largely been caused by the impending change of the until recently predominant scientific (rationalist) paradigm in first generation cognitive science (Mey 1982; Chomsky 2002), which is being slowly 'squeezed' out by the *non-Cartesian paradigm* that characterizes second generation cognitive science.

A major problem cognitive linguistics sets out to resolve is how knowledge is represented (manifested) in language. This sets specific guidelines for contemporary cognitively oriented research, such as the exploration of essential functional properties of signs as empirical objects inseparably connected with human subjects (Kravchenko 2003a). That signs are subjective by nature was rather unambiguously indicated by Peirce, and the idea was taken further by Morris. Placing the issue of the nature of sign in the context of *semiosis*, he argued for an additional fourth factor to be included, alongside with Peirce's triad of the *sign form as such* ('Representamen'), *designatum* (Object) and *Interpretant*, namely, the *Interpreter*. Morris emphasized that

> the terms 'sign', 'designatum', 'interpretant', and 'interpreter' involve one another, since they are simply ways of referring to aspects of the process of semiosis. Objects need not be referred to by signs, but there are no designata unless there is such reference; something is a sign only because it is interpreted as a sign of something by some interpreter; a taking-account-of-something is an interpretant only in so far as it is evoked by something functioning as a sign; an object is an interpreter only as it mediately takes account of something. The properties of being a sign, a designatum, an interpreter, or an interpretant are relational properties which things take on by participating in the functional process of semiosis. Semiotic, then, is not

concerned with the study of a particular kind of object, but with ordinary objects in so far (and only in so far) as they participate in semiosis (Morris 1945: 55).

This means that any random perceivable object can become a sign as long as in the mind of the perceiver it stands in certain relations with other objects (phenomena) becoming part of a network of specific associations (Doroszewski 1970).

Morris' programmatic work *Foundations of the Theory of Signs* took the science of semiotics a long step further down the road laid by Peirce. However, in the decades that followed the publication of his ingenious work the emphasis laid by Morris on *unification of science* as the goal of his semiotic program somehow dissipated in the numerous discussions of his theory of signs, and the role he attributed to semiotics as a science among other sciences is only beginning to be appreciated by modern science:

> Semiotics has a double relation to the sciences: it is both a science among sciences and an instrument of the sciences. The significance of semiotics as a science lies in the fact that it is a step in the unification of science, since it supplies the foundations for any special science of signs, such as linguistics, logic, mathematics, rhetoric, and (to some extent at least) aesthetics. The concept of sign may prove to be of importance in the unification of the social, psychological, and humanistic sciences in so far as these are distinguished from the physical and biological sciences. And since it will be shown that signs are simply objects studied by the biological and physical sciences related in certain complex functional processes, any such unification of the formal sciences on the one hand, and the social, psychological, and humanistic sciences on the other, would provide relevant material for the unification of these two sets of sciences with the physical and biological sciences. Semiotics may thus be of importance in a program for the unification of science, though the exact nature and extent of this importance is yet to be determined (Morris 1945: 2).

The cited passage gives a fairly good idea of what Morris' theory of signs really is inasmuch as its scientific import is evaluated: it is a necessary theoretical prerequisite for the working out of a unified science, or metascience:

> ...It is possible to include without remainder the study of science under the study of the language of science, since the study of that language involves not merely the study of its formal structure but its relation to objects designated and to the persons who use it. [...] But since nothing can be studied without signs denoting the objects in the field to be studied, a study of the language of science itself must make use of signs referring to signs, so semiotic must supply the relevant signs and principles for carrying on this study. Semiotic supplies a general language applicable to any special language or sign, and so applicable to the language of science and specific signs which are used in science (*ibid.*, p. 62-63).

The core message of *Foundations* seems to have been largely lost on linguists and philosophers many of whom continue to overlook its methodological implications: "Problems which are often classed as epistemological or methodological fall in large part under semiotic" (*ibid.*, p.56).

That the nature of sign cannot be understood apart from the study of the nature of meaning, just as the problem of meaning cannot be satisfactorily resolved without the nature of sign entities being considered, is hardly a matter of dispute today: both are inseparably connected by the main function of language. It seems appropriate here to recall the main law of semiotics with its formula "a sign for a sign" first suggested by Peirce (although the principle itself was understood by Leibnitz): a sign may be interpreted by another sign or a string of signs, and that this is the only way for a sign to develop its meaning. It implies that the semiotic process has, essentially, a circular organization, just like the process of the living.

To understand the function of language one must assume its inherently biological character, and since language is only one of the many biological features possessed by humans as living organisms, scientific study of language is impossible without the study of man as a living system. As Ronald Kellogg (1994: 5) observes, "the survival of new meanings depends on competition and successful adaptation in a manner that parallels the survival of genes in biological evolution". Thus we realize that if we want such a study to be fruitful, we must contend that it requires a biological approach to linguistic phenomena, particularly, to meaning (signification) as the root of semiosis. This realization gave rise to *biosemiotics* as the study of signs, of communication, and of information in living systems. If, to use the traditional concept of 'information', language is viewed as an information system and compared with the information system of the human genome, a striking isomorphism is revealed, "raising the phenomenological issue of the nature of [these] systems and of the causes that led to such structural isomorphism" (Gamkrelidze 1988: 6).

2.5.2. Biosemiotics

Biosemiotics is not one among other biological sub-disciplines but rather constitutes a certain theoretical frame for the study of biology when all processes going on in animate nature at whatever level are analyzed and conceptualized in terms of their character of being sign-processes (Bouissac 1998). In other words, the subject of biosemiotics is "investigation of the biological nature of signs and semiotic base of biology" (Sharov 1992: 345). As a matter of fact, the idea that semiotics and biology have very much in common was present in Ernst Cassirer's (1923) famous definition of man as the "symbolical animal", so it is not altogether new. What biosemiotics does, however, is take the concept of symbolical animal further and apply the concept of sign-process (semiosis) to vegetative processes (phytosemiotics) and animal behavior (zoosemiotics) (Sebeok 1963).

Speaking of how biosemiotics differs from other kinds of semiotics, Claus Emmeche (1999) suggests that one should distinguish between anthroposemiotics, biosemiotics, and pansemiotics as three answers to the question of what a science of signs should deal with, namely, that sign production and sign interpretation is just a human phenomenon (Saussure), or a general biologic process (Morris), or a universal physical phenomenon (Peirce). He notes, however, that anthroposemiotics may also be defined as the branch of general semiotics that deals with human communication systems, while biosemiotics deals with all other communication systems and includes zoosemiotics, phytosemiotics, endosemiotics, etc., i.e. there is no basic conflict between *biosemiotics* and *anthroposemiotics*.

The term *biosemiotics* was first used by Friedrich Rothshield (1962) who suggested such an approach to the study of self and personality which presupposed that the history of subjectivity did not start with man, but that the human spirit was preceded by many preliminary stages in the evolution of animals. About ten years later the term was used by the Russian linguist Juri Stepanov (1971) to refer to the branch of science that deals with sign systems of living beings; however, for quite some time biologists in other countries whose research was similarly directed remained unaware of this novel approach, a circumstance which was only far too common in science before the advent of the World Wide Web. The term 'biosemiotics' finally appeared in international literature in mid-80s, introduced by Thomas Sebeok, and as a school of semiotic thought biosemiotics began to take shape only in the 90s. Essentially, it is a logical evolutionary continuation of the direction in the study of signs set by Charles Peirce a hundred years earlier (Braga 1999) and continued by Charles Morris (Petrilli 1999).

Thomas Sebeok (1979) picked up Peirce's and Morris' methodological lead by suggesting that a full understanding of the dynamics of semiosis may in the last analysis turn out to be no less than the definition of life, and set out to bring together scholars to whom this idea did not seem altogether alien (Sebeok and Umiker-Sebeok 1992). In a special issue of *Semiotica* (Sebeok et al. 1999) a few years later it was noted that biology and semiotics traditionally belong to opposite scientific camps in the divided conception of the world we have inherited from the Cartesian tradition. One is supposed to deal with living systems in terms of matter and energy, the other with mind-processes of signification and communication. However, in recent decades it has become increasingly clear that this division can no longer be sustained. Organisms are inherently informational and communicational structures (Barbieri 2006), and minds are corporeally embodied. All true signs in the world are, ultimately, dependent on biological organisms, whether for communication, control, or construction, and whether at a cellular, organismic or social level. We cannot understand signs fully until we understand their role in the organization of life. So, in the biosemiotic conception the life sphere is permeated by semiosis and signification: whatever an organism senses also means something to it, and all organisms are thus born into a *semiosphere*, i.e., a world of meaning and communication (Bouissac 1998).

In view of this, biosemiotics may be defined as an integrative science of these two knowledge areas (living systems and signification/communication) aiming at working out a coherent theory of semiosis of life and life as semiosis[4]. This exciting development has become possible thanks to the advances in modern biology:

> With the discovery that a set of symbols has been used by nature to encode the information for the construction and maintenance of all living things, semiotics — the analysis of languages and texts as sets of signs and symbols — has become relevant to molecular biology. Semiotics has given students of the DNA text a new eye for reading, allowing us to argue for the validity of a multiplicity of meanings, or even for the absence of any meaning, in a stretch of the human genome (Pollack 1994: 12)

This 'new eye' allowed biologists to see dual coding in *all* living systems represented by the pair DNA–cell (or genome-phenotype) (Hoffmeyer and Emmeche 1991). Consequently, and in agreement with our intuitions, observes Robert Pollack (1994: 153),

> we can expect to find more and more examples of the richness of a real language in our cells. DNA and protein have grammar and syntax, and we have already come upon typographical errors, double meanings, synonyms, and other subtleties.

DNA stores knowledge which remains stable over several generations, it serves as a long-term memory-bank for the organism so that knowledge already compiled is not forgotten too easily. Yet organisms must also adapt themselves to more rapid environmental changes, in time-scales shorter than their lifetimes, and the only possibility for individual adaptation within an individually specific environment is in phenotypic adaptations. Environmental knowledge of an organism must not only be *stored* in a stable form (in genes), but there must also be mechanisms to *update* it (Plotkin 1987). According to Tommi Vehkavaara (1998: 214),

> knowledge is stored in some *structure* as its *order*. The material of a structure is not essential, and it is not required to be linguistic or mental — DNA, cytoskeleton, neural network, or social practice may suit as well. However, this structure has to be accessible or interpretable to the knower — quite often the structure is part of the subject, but not necessarily (as is the case with writing). It is the *order* of a

---

[4] For a short review of the history of trends and ideas in biosemiotics in the 20th century see Kull (1999).

structure that represents the object — the order or information that can be copied and transformed into other structures.

This kind of knowledge is identified by Vehkavaara as *structural* knowledge, which, he admits, may also be stored in linguistic structures. It is interesting to note here the direct conceptual and terminological coincidence with the ideas about the distinction between *structural* and *phenomenological* knowledge expressed and developed earlier in linguistics (Goldsmith and Woisetschlaeger 1982; Kravchenko 1992; 1996). Although Vehkavaara makes no mention of *phenomenological* knowledge as a counterpart of structural knowledge, it clearly correlates with the phenotype member of the pair DNA–cell, because it is cells that act as interpreters of structural knowledge, and this interpretation depends on an individual cell's environment (niche).

The biosemiotic conception of knowledge and its organization appears thus to be not unlike one in cognitive linguistics. And this is not surprising, because "the life science and the sign science... mutually imply one another" (Sebeok 1994: 114). As Aleksei Sharov (1998: 404-405) observes,

> sign processes penetrate the entire body of an organism. [...] Signification is the fundamental property of living systems that can be taken as a definition of life. Hence, biosemitics can be viewed as a root of both biology and semiotics rather than a branch of semiotics.

In other words, life may be defined as *functional interpretation of signs in self-organized material code-systems making their own Umwelts* (Emmeche 1998). Such definition of life contains three interrelated notions which are very important for its adequate understanding: *self-organization* as a distinctive feature of a living system, *materiality* of this system viewed as a particular order of matter-energy configurations which is a system of knowledge and molecular organization as well, and *Umwelt* as the subjective world of an organism as a living system. In biocognitive philosophy of language, the notion of self-organization and the principle of circularity in the organization of the living entailed by it, are indispensable in understanding the organization and function of language as a natural (biological) sign system. The notion of the materiality of sign phenomena allows to depart from Cartesian dualism in understanding signification as some mental or intentional content, which is a substantially different phenomenon from being a sheer physical thing. Instead, signification may be understood as something arising from an organism's interactions with its environmental niche perceived as the organism's subjective (that is, construed by the organism as a result of such interactions) world, or *Umwelt.* The notion of *Umwelt* introduced by Jakob von Uexküll (1909) appears to be very similar to the notion of *niche* which belongs to the set of concepts used in autopoiesis and which is essential in understanding the origins of signification and the meaning of signs.

The latest developments in biosemiotics show a clear epistemological turn in approaching biological problems, when a certain philosophical basis for biology is proposed which introduces subjectivity (i.e., when an organism is viewed as a subject) and helps understand the development of mental features through the semiotically interpreted evolutionary epistemology (Schult 1989; Hoffmeyer 1996). One might say that Morris' dream to see unification of science in semiotics is finally beginning to materialize in the unification of biology which, according to Hoffmeyer (1997), has to be based on the fundamentally semiotic nature of life.

2.5.3. A Biocultural Theory of Meaning

This understanding is characteristic of a unified biocultural theory of meaning proposed by Jordan Zlatev (2003). In a nutshell, it is an integrative theory of meaning based on the concept of *value*, understood both as a biological and as a sociocultural category, synthesizing ideas from evolutionary and developmental psychology, semiotics, and cybernetics. One of the basic tenets of the theory is that only living systems have intrinsic value which is a necessary and sufficient condition for possessing the category *meaning*. Therefore, the theory refutes positive assessment of the meaning potential of (current) artificial systems.

As Zlatev observes, the conception of *meaning* has become increasingly fragmented and different kinds of meaning theories are often separated by deep trenches. Yet, despite its very controversial character, meaning still remains a fundamental concept in philosophy and linguistics, and extreme fragmentation of views on its nature is hardly defensible. Therefore, there is a call for an *integrative* conception of meaning. Such integrative conception in a very condensed form is described as follows:

Meaning (M) is the relation between an **organism (O)** and its physical and cultural **environment (E)**, determined by the **value (V) of E for O.**

$$\mathbf{M = V\ (O, E)}$$

The unified theory of meaning is defined in terms of six theses:

(1) Meaning is the relationship between an *organism* and its *environment*, determined by the *value* which particular environmental *aspects* (falling into *categories*) hold for that organism.

(2) The meaning of physical aspects, perceived via *innate value systems*, is based on their role for the preservation of the life of the individual and its kin.

(3) The meaning of cultural aspects is evaluated on the basis of *conventional value systems* consisting (predominantly) of *signs*. These need to be acquired before they can determine the meaning relation for the organism.

(4) Both innate and acquired value systems serve as *control systems* by directing and evaluating the organism's behavior and its adaptation.

(5) Intrinsic value, and consequently meaning, is intimately connected to *emotion* and *feeling* and represents a necessary, but not sufficient condition for phenomenal experience.

(6) On the basis of the concepts introduced in (1) – (5), four different types of meaning systems can be defined, forming an *evolutionary and epigenetic hierarchy* (*cue-based, associational, mimetic* and *symbolic* meaning systems). These meaning systems, viewed from different angles, are presented by Jordan Zlatev in Tables 1, 2, and 3.

Table 1. A hierarchy of meaning systems based on value, defined by (a) nature of the value system and *level-specific* (b) structures of environmental categorization, (c) structures of social interaction, (d) type of learning mechanism and (e) more speculatively, structures of phenomenal experience, i.e. consciousness

Meaning system	**Value system (VS)**	**Categorization**	**Social interaction**	**Learning**	**Emotions and consciousness**
Cue-based	Innate VS	Fixed cues	—	Habituation	—
Associational (including Cue-based)	Innate VS + acquired associations	Associations Image/action schemas	Innate and partially learned signals	Conditioning	Primary emotions
Mimetic (including Associational)	Innate VS + acquired associations + mimetic conventions	Mimetic schemas	Gestures Mimes	Imitation learning	Secondary emotions Intersubjectivity
Symbolic (including Mimetic)	Innate VS + acquired associations + mimetic conventions + symbols	Classes Hierarchies Stories	Symbols Propositions Narratives	Reflection	Higher-order consciousness

Table 2. The phylogeny of meaning; creatures with four types of meaning systems: cue-based, associational, mimetic and symbolic

Meaning system	**Which creatures?**	**Neural structures**	**Characteristics**	**Limitations**
Cue-based	From bacteria to flatworms	Elementary nervous system	Homeostasis Pre-determined (non-phenomenal) *Umwelt*	(Almost) no learning No emotion
Associational	From invertebrates to most mammals	Central nervous system, (in birds and mammals: limbic system, cortex)	Sensorimotor schemas Signals	No general learning No imitation No intraspecies variation, 'culture'
Mimetic	Apes (and cetaceans)	Developed "mirror neuron system" (Rizzolatti and Arbib)	Imitation learning Intraspecies variation, 'culture' (Some) reflection	No cultural history No language No full 'theory of mind'
Symbolic	Homo sapiens = "Homo symbolicus"	Extended prefrontal cortex (Deacon)	Language Cumulative culture Reflection	'Ungrounding' from somatic value Self-destructivity

Table 3. The ontogeny of meaning; four developmental stages: cue-based, associational, mimetic and symbolic

Developmental stage	**Which age?**	**Characteristics**	**Limitations**
Cue-based	Pre-birth	Reactive, predetermined behavior	No value-based learning
Associational	0-9 months	Ontogenetic ritualization Protoconversation Humor	No true imitation No understanding of intentionality
Mimetic	9-24 months	Joint attention Imitation (.Secondary) intersubjectivity Declarative pointing Protolanguage	No full conventionality No true language No narrative
Early symbolic	2-5 years	Vocabulary spurt Grammar spurt Emergence of narrative	No full 'theory of mind' Limited meta-cogmtion

Using his integrative theory of meaning, Zlatev addresses the classical question, "What, if any, kind of artificial system would be capable of meaning?" He argues for two conditions for any system, natural or artificial, to have meaning:

> (1) It must have an internal value system controlling its perception, behavior and learning *autonomously.*
>
> (2) This value system must be *intrinsic* to the system in the sense that it serves to preserve the system's *organization,* rather than criteria which are external to the system (defined by the system's designers in the case of artificial systems).

While many cybernetic systems fulfill, to various degrees, condition (1), there is no current artificial system that satisfies condition (2), while any living system does so. Hence, concludes Zlatev (obviously supporting Searle's (1984) earlier categorical argument against strong AI), "there are no artificial meaning systems, even of the simplest kinds, and *meaning is co-extensional with life!*" (Zlatev 2003: 282).

On comparison with what has been discussed in the previous sections, the synthetic character of the biocultural theory of meaning is obvious. It leans conceptually on autopoiesis and biosemiotics, with a cultural aspect of life (not unlike the one discussed by Tomasello and Rakoczy (2003)) as an added dimension, in its construal of meaning as a cognitive model for successfully functioning living systems. Because of this it seems to be a very promising theoretical framework as it opens a new horizon in the study of meaning as the key concept in cognitive science and semiotics whose ultimate goal is to understand the essence of human life in the context of its physical and social environment. Besides being a unified theory and comprising semiotics, biology, sociology, evolutionary psychology and cybernetics, it appears to be an evolutionary follow up on Peirce's and Morris' ideas of the unity of science embodied in sign systems and is probably one of the first methodologically consistent attempts at a holistic conception of human language as a natural semiotic/cognitive phenomenon.

2.6. Distributed Language and Cognition

Today, the conceptual framework of mainstream linguistics is still based on the assumption that linguistic signs represent (denote) different aspects of reality, thus denotation is seen as the basis for understanding in human communication. However, the representational theory of mind built largely on this notion and implying a kind of non-arbitrary (i. e., computationally definable) relationship between the sign and what it stands for has been unable to facilitate radical advances in areas where application of the coded equivalence principle should do the job, such as machine translation or Artificial Intelligence. But if a theory fails to do the job it has been devised for, something must be wrong with the theory itself or at least with its grounding principles. It is not surprising that linguistics as a science increasingly shows signs of a profound methodological crisis, and, to quote Teun van Dijk (2006: 7), young scholars in their research "feel torn between the vast amount of different theories, methods, approaches, schools, sects and gurus". There is even talk of linguistics as 'voodoo science' (A. Gross 2005).

As emphasized by Nancy Ritter (2005), there is an apparent need to integrate results from various avenues of research. Teun van Dijk (2006: 7) sides with her in observing that "originality and renewal typically take place at the boundary of different disciplines or approaches or by combining different theories and methods". Yet a combination of different theories and methods in the study of language and communication of itself cannot guarantee imminent progress unless the old Cartesian epistemology continues to define the tools and procedures used in the application of these theories and methods. There is a call for a new non-Cartesian epistemology that would help overcome the crisis caused by dualistic treatment of language and communication. As I have argued elsewhere (Kravchenko 2003a), such epistemology is offered by autopoiesis as the theory of the living. The autopoietic framework allows for a holistic approach whereby language is viewed as adaptive functional activity based on an organism's experience of the environment to which the organism stands in a relation of reciptocal causality.

As an alternative to orthodox linguistics, the *distributed view of language and cognition* gives an account of language as a specific kind of cognitive activity integrated in human interactions with the environment. Experimental data highlight the role played by experience in perception, categorization, and conceptualization. It has been shown that background knowledge affects categorial decisions (Palmeri and Blalock 2000; Gelman and Bloom 2000) and acquisition of new concepts (Nelson et al. 2000; Matan and Carey 2001); that meaning is specifically related to perception (Allwood and Gärdenfors 1999), which itself is influenced by categorization processes (Schyns 1997; Albertazzi 2000), and that object recognition and categorization is largely an on-going process affected by the experience of our environment (Wallis and Bülthoff 1999). Empirical data suggest that, contrary to the belief of first generation cognitive scientists,[5] brains do not use unchanging quasi-linguistic representations. Connectionist models and neuroimaging show neural activity to be radically distributed: the 'representational states' of a brain are time-dependent. Moreover, animals and robots can often perform flexibly and adaptively without making use of inner representations. Accordingly, many cognitive scientists now focus on situatedness, embodiment and pattern-matching. To the extent that it makes sense to invoke inner representational states at all, these are distributed within the brain, based on a history of body-world coupling, and unlikely to recur in identical form. To explain the importance of verbal patterns in human life, therefore, we must also rely on cultural evolution. For this reason a growing number of researchers reject 'internalism' by arguing that what we call 'mind' relies on causal processes that spread beyond skin and skull (e.g. Vygotsky, Gibson, Maturana and Varela, Clark, Hutchins, Järvilehto).

Distributed view of language and cognition is based on the assumption that far from 'representing' cognitive processes, language transforms the causal links between brain, body and world (see A. Clark 1997). Language describes a heterogeneous set of artifacts and practices which enable us to exploit behavioral modalities in ways allowing the attribution of semiotic values. For example, in contrast to the classical view on language and cognition, the distributed view takes it as

[5] While many positions are possible, the classic view drew on serial digital computers to regard representations as quasi-linguistic or static entities whose determinate formal properties remain largely unchanging between persons.

axiomatic that the semiotic value of a sign is not given in advance but emerges as a function of integrating activities (and in experiential time), or that aspects of the environment determine, at least, the content of cognitive processes, and knowing and coming to know do not depend exclusively on events within the skin — rather, these activities are inseparable from acting in relation to perceived aspects of an environment (Frisina 2002).

This view is consistent with the main tenets of the biological theory of cognition (Maturana 1970; Maturana and Varela 1980; Järvilehto 1998), radical constructivism (Glasersfeld 1995), biosemiotics (Sebeok and Umiker-Sebeok 1992), integrational linguistics (Harris 1996) and biocognitive philosophy of language (Kravchenko 2003a; 2006). There has been a growth of converging evidence from different areas of research in humanistic sciences highlighting the ineptitude of mainstream orthodox linguistics to deal with the problems it has been called upon to resolve because of its initial conceptual fallacies. The paradigmatic shift that we are witnessing today toward biologism in explaining language and cognition may help find a way to get out of the methodological deadlock and signal a new era in language and cognition research. However, there is strong opposition to post-computational thinking about language and cognition. In the introduction to the special issue of *Pragmatics and Cognition* on distributed cognition its editors Stevan Harnad and Itiel Dror (2006) outline their stance toward the dynamics of cognition arguing against the distributed view of cognition. Their brief exposition is, in fact, a follow-up on Harnad's (2005a) more extended treatment of the concept of distributed cognition, where it was claimed that "there is no such thing as distributed cognition, only collaborative cognition" (p. 501). According to Harnad and Dror, cognition is

(i) autonomous activity (internal process) which "takes place entirely within the brains of cognizers",

(ii) this activity generates input/output capacity which is what "higher (and possibly) lower vertebrates are able to do in their world", and

(iii) in the case of the human species, cognition is the ability to pass the Turing Test (TT) which is "an audition for cognition".

This is what is known as the internalist (computational) account of cognition (cf.: Chomsky 1975; Ajjanagadde and Shastri 1991; Pinker 1995b; Fodor 1998 *inter alia*) based on the metaphor 'thinking is computation' borrowed by Turing from Hobbes's computational theory of mind (Dascal 2006). I rather like Harnad and Dror's brusque style, but I find it hard if not impossible to share their assumptions about cognition.

I have great respect for Alan Turing even though for one simple reason: few thinkers have been able to cast such a strong spell on the minds of other 'professional cognizers' blinding them to any empirical evidence and/or theoretical arguments against the infamous claim that 'thinking is computation'. Harnad and Dror present a classical example of Cartesian position with their ultimately reductionist (vitalistic) approach to cognition viewed as a function of the brain rather than the function of the living system as a whole. Considering recent developments in cognitive science, which is by no means conceptually limited to symbolist models of language and cognition (i. e., as computations on abstract symbols found in the head), it really makes one wonder.

If 'cognition is what cognizers do' , what exactly is it that they do, and why do they do it?

Given that cognizers are organisms, they obviously do what they do to survive and reproduce. So, basically, we are speaking of organisms' interactions with the environment in which they must survive and reproduce, and all organisms which successfully interact with the environment to this end are cognizers regardless of the level of their organization — from amoebae to humans. Yet, according to Harnad and Dror (2006: 209), 'cognition takes place entirely within the brains of cognizers', and if 'cognizing is whatever goes on inside the brains of organisms to give rise to their performance capacity, then cognition is the functional substrate — the generator — of their know-how'. Here arises the first contradiction in their position: if cognizers are organisms, and 'cognition takes place entirely within the brains of cognizers', then how are invertebrates and other brainless organisms (such as of the phylum *Protozoa*) capable of cognition? Surely one cannot deny that even the amoeba displays what may be described in Harnad and Dror's terms as basic know-how functional competence which, ultimately, enables it to survive. As noted by Peter Harder (1999), the problem with mainstream cognitive science is that the prototype of cognition that pervades first generation cognitivism are higher-level processes, specifically those which involve symbolic-representational content.

By equating cognition with whatever goes on inside the brain, Harnad and Dror leave open the question of how sensorimotor activity and perception enter the picture. As Clark (1997: 221) rightly observes, "the traditional divisions among perception, cognition, and action look increasingly unhelpful". True, elsewhere Harnad (2005b), paying tribute to Gibson, admits the importance of sensorimotor systems, but the way he does it ("the things in the world come in contact with our sensory surfaces, and we interact with them based on what that sensorimotor contact 'affords'", p. 20) suggests that organisms that can't be included in the extension of the indexical *we* (= "we, humans") deserve no attention as cognizing agents. I am willing to concede that Stevan Harnad doesn't really think so, and the question arises simply from his somewhat inaccurate use of language — as when he says 'our sensory surfaces' meaning, probably, 'the sensory surfaces of any and all organisms'. I am also inclined to think that by confining cognition to the brain

Harnad and Dror do not refer to cognition understood broadly as an organism's successful interactions with the environment, but to a special kind of 'minded' cognition characteristic of higher order organisms, such as vertebrates, which do have brains. But if this is the case (which it surely seems to be), their definition of cognition as a process should be reformulated into something like 'minded cognizing is whatever goes on inside the brains of organisms' and contrasted to 'mindless cognizing' as, for example, in the case of *Protozoa*. And this would inevitably bring us to the time-old problem of mind with its central issue of the relationship between mind and body on the one hand, and mind and cognition on the other, not to mention the language/mind relationship and its crucial role in drawing a divide between minded cognizing in humans as different from minded cognizing in other higher order vertebrates. Harnad and Dror's tacit assumption that cognition (or should we say, "mind"?) is the functional property of the brain which exists in space and time and has its locus in the head, leads to another contradiction rooted in their definition of cognition: on the one hand, 'cognition is what cognizers do', that is, interaction with the environment, yet on the other hand, it 'takes place entirely within the brain'. This contradiction can be reconciled only on condition that 'whatever goes on inside the brains of organisms to give rise to their performance capacity' is caused ('triggered') by the environmental cues and thus cannot be viewed as an autonomous capacity generated solely by the internal processes, but rather as part of the complex organism-environment system when purposeful behavior is explained by joining an organism's *phenomenal world* (the world as perceived) and its *effector world* (the world as enacted) into a single closed whole (Uexküll 1909; Järvilehto 1998). And this is precisely what Harnad and Dror reject in their argument for the autonomy of cognition.

The entire concept of cognition based on the computational metaphor according to which a human being is an autonomous system with an input/output capacity, depends crucially on the afore mentioned divisions — despite heaps of empirical evidence to the contrary. Surely Bennett and Hacker (2002: 432) are right when they say: "The suggestion that we should think of ourselves as computer programs is not coherent", because the brain is simply not a computer (Lamb 1999). This realization has caused a shift in humanistic science toward biology and its integration with other disciplines such as cognitive philosophy, psychology, semiotics, linguistics, etc. Central to these processes is the realization that *cognition is a biological phenomenon* and it can be understood only as such; any understanding of the cognitive domain from the point of view of epistemology involves understanding it from the point of view of biology. To attain such understanding, two questions must be posed (which Harnad and Dror in fact do): *What is cognition as a function?* And *What is cognition as a process?* Much depends on how these questions are answered, and the answers offered by Maturana's *biologi-*

cal theory of cognition lead to a very different stance toward cognition than Harnad and Dror's.

On the biological account of cognition, the cognizer's 'autonomy' advocated for by Harnad and Dror, is a *relative* notion that bears on the observer's cognitive domain which, by definition, does not coincide with the cognitive domain of the observed. True, in the biological theory of cognition we speak of the autonomy of an organism as an autopoietic system preconditioned by autopoietic closure which is realized through a continuous structural change under conditions of continuous material interchange with the medium. This is the metabolic or physiological domain. But in the relational or behavioral domain in which it interacts and relates with the medium that contains it as a totality, external factors play a significant role in determining cognizers' mental states as a result of the coupling between the environment and the organism; this coupled system is one in which brain, body, and world move in and out of reciprocal causal relations. As argued by Lynn Margulis (1989), the so-called scientific view that life adapts to a passive environment of the physical world is severely distorted. As it is, life makes and forms and changes the environment to which it adapts, and then that environment in its turn feeds back on the life that is changing and acting and growing in it. This understanding allowed Kurt Lewin (1946) to speak about cognition in terms of a 'life field' when "the person and his environment have to be considered as *one* constellation of interdependent factors" (p. 793).

We human beings, argues Maturana (1988), assess cognition in any domain by asking a question and demanding an explanation. If what we observe as an answer satisfies us as adequate behavior or as adequate action in the domain specified by the question, we accept it as an expression of cognition in that domain, and claim that he who answers our query knows. This is, basically, the ideology behind the TT. Yet from the point of view of biology, knowing has to do with the congruent interactions of entities that are structurally determined systems in which all that happens with them and to them is determined at every instant by the way they are made (their structure) at that instant. To describe the observed complex dynamics of this system as something 'done' by cognizers is misleading since it is not so much the question of *how cognizers do* what we think they do when we observe their behavior, but the question of *what happens* in the cognizer as a living system when we observe its behavior. Brains do not 'generate' behavior, what we call "behavior" is our convenient way of referring to the relational domain of interactions of the observed organism with its niche, while such interactions are the outcome of a history of structural coupling of that organism as a structure determined system with the environment. Thus, it is only natural to ask, "Is the Turing Test really 'an audition for cognition'"?

According to Harnad and Dror (2006: 210), the purpose of the TT was to "design a machine that could pass it, in order that we should at last understand what cognition is, what generates the know-how of an autonomous system like us". In the TT, to qualify as a cognizer "the candidate must be able to exhibit the performance capacities *of* a normal human being, indistinguishably *from* a human being, *to* a human being" (*ibidem*). What it means is that a robotic system should be able to display interactional behavior which might be interpreted by the human observer as a kind of behavior he, the human observer, would display in a similar interactional situation (this is what we call 'understanding'). But we all know that appearances are deceitful, and what a human being *thinks* he sees is not necessarily *what* he sees. As Ezequiel Di Paolo (2005: 446) argues,

> [t]he movement of meaningful action can be convincingly emulated in an artificial system but this is not the same as the system acting meaningfully. The robot may look scared and retreat when yelled at but this may be only a sophisticated illusion; there is no way to tell just from observing its responses, Turing-test style. Being functionally scared is not the same as being scared.

In the TT, the observed behavior displayed by the robot is viewed as an output capacity generated by an autonomous cognizing agent in response to some input, and cognition is the internal processes responsible for the generation of the output. So, to understand cognition, we must answer the question, "How is the input/output capacity generated?" And this is the third problem with Harnad and Dror's account of cognition which has to do with how they answer the question, "What is cognition as a process?"

When Harnad and Dror speak of the necessity to understand cognition as what goes on in the brain of an individual cognizer, what they really mean is to understand the mind. By tacitly assuming that a robot which can successfully pass the TT has a mind, they commit the classical Cartesian error of divorcing mind from the body. Moreover, attempts to understand 'minded' cognition in an artificial system such as a robot which, as an entity, came into existence as a result of 'minded' behavior (interactional output) of a living system while this minded behavior of the latter remains a mystery, are doomed from the start. Yes, argue Maturana, Mpodozis and Letellier (1995), in our culture we do speak of mind to explain phenomena that the observer distinguishes as taking place in the relational space of the organism: intentions, purposes, concerns, etc., and we speak as if we were referring to an entity that may have a location in the brain and may interact with other minds or the body. But from the point of view of cognition understood as a biological function (as briefly outlined above), it should be apparent that there is no such thing as 'the mind' in the operation of the nervous system, and that 'the mind' is nothing but an explanatory notion. "Language, self-consciousness and mindedness are different forms of existing in the relational domain in which a living being lives, not manners of operation of the nervous system" (*ibid.*, p. 25). In these circumstances, the questions about mind/body interactions can be reformulated as follows: 'How do relational phenomena have consequences in the body dynamics?'

Because of the terms in which the Turing Test is defined, it can be more or less reliably applied when (i) it is generally agreed what makes a 'normal human being', (ii) the observer (the tester) is a normal human being. I suspect that human cognizers are chosen as best testers for the TT because of the convenience rendered by their linguistic ability. As was mentioned above, we assess cognition by specifying the domain with a question and demanding adequate behavior in that domain. But language makes human cognition so profoundly different from animal cognition as to make it virtually impossible to run the TT on animals for obvious reasons.[6] Harnad and Dror might object that the TT is a robotic one, including both a person's linguistic capacity and the sensorimotor capacity in which it is grounded. Here, again, we have a conceptual problem that affects the perspective they take on the TT and the language they use to describe it. In the biology of cognition, linguistic behavior is not an input/output capacity, it is behavior in a consensual domain and as such it is an additional dimension of the cognitive domain of orientational interactions. This domain is established by interactions of the orienting organism and the organism to be oriented with representations as states of relative neuronal activity. In other words, language is not a capacity but a *relational domain.* For a robotic cognizer to enter this relational domain (and for relational phenomena to have consequences in its body dynamics), its domain of interactions and the domain of interactions of the human observer (that is, their cognitive domains) must coincide to a large degree. But in the biology of cognition where cognition is co-extensional with life, a cognitive system is a system whose organization determines the domain of interactions in which it can act meaningfully to sustain itself. There can, therefore, be no sensible talk of a robot possessing the properties of a living system unless it can sustain itself in the metabolic or physiological domain in which its components realize it as an autopoietic entity. As of now, such robots do not exist.

Language dynamics are processes of using and interpreting language as a person engages with the environment. These depend on the causal processes that constitute cognitive dynamics which occur in and across several time domains — e.g. evolution, history, development, relationships, experiential time, various microdomains. A normal human being is a languaging human being, it is a human being whose existence is specified by linguistic interactions with others in a consensual domain. If a normal human being is a languaging human being, then it is also a

[6] One such reason, for example, is our inability to read an animal's 'mind' because, being structure determined systems which exist in different consensual domains (the relational or behavioral domains in which animals and humans interact and relate with the medium that contains them as totalities), we cannot possibly enter the relational domain of an animal, nor can an animal enter the relational domain of a human being.

social cognizer whose observed behavior is a description of its interactions in a consensual domain. It appears, in a rather obvious way, that the concept of 'normal human being' cannot be divorced from what goes on in the 'minds' of other human beings and their linguistic behavior which is a description of such goings-on that produces orientational effects on the observer. Yet such orientational effects which can have any meaning to the observer at all are possible only if the observer has shared experience with the observed in a consensual domain, because brains are *dialogical* (Linell 1998). In other words, for the Turing Test to meet the necessary conditions, shared experience in a consensual domain of linguistic interactions of the observer and the observed as social cognizers must be factored in, which is to say that the original purpose of the TT ('to understand what cognition is by designing a machine that could pass it') *is simply non-implementable* as long as contemporary androids remain what they are — lifeless machines capable of *simulating* at least some cognitive processes but incapable of languaging in a consensual domain and thus incapable of human cognition which, as a life process, is *distributed.* The TT will be passed not when we can 'generate' the capacity to pass it, but when we can (if at all) create a *bona fide* cognitive system — a living organism.

2.7. Summary

As I have tried to argue, the direction in which today's cognitive linguistics (and, by extension, cognitive science) is going was set out by its precursors in philosophy, semiotics, and biology, above all — by the founders of autopoiesis as the biological theory of cognition. However, this theory has not received the full attention it deserves as it is built on epistemological foundations different from those shared by the mainstream cognitive science. One of the main reasons why it should be so has been explicitly formulated by Margaret Boden (2000: 132-133):

> The vocabulary of autopoiesis is unfamiliar to most biologists... [...] On the one hand, they reject theoretical terminology that is common in both biology and psychology. On the other hand, they speak of knowledge and cognition in many contexts where psychologists and most cognitive scientists would not.

Yet, to borrow a vivid (although slightly modified) expression from Searle (1984), the difficulty is that most cognitive scientists persist in talking about a twenty-first-century problem (the nature and essence of cognition as a biological phenomenon) in an outmoded mid-twentieth-century vocabulary of the first generation cognitive science. As an alternative theory of life and/as cognition autopoiesis seems to be much more promising as it enables to make linguistics *de facto* rather than *de jure* a true science with an empirical object of study. For this to happen, the idea of cognition should be rid of the pernicious computational metaphor and its blinding effect on the development of cognitive science, particularly in the field of Artificial Intelligence research.

The turn-of-the-century empirically oriented research unambiguously indicates that cognitive science is in the process of working out a more comprehensive, that is, holistic picture of language and cognition. A new philosophy of language is emerging, a bio-cognitive philosophy that treats language as *adaptive functional activity based on an organism's experience of the environment (the world) to which the organism stands in a relation of reciprocal causality.* If natural spoken language is treated as a kind of adaptive behavior of an organism (a description of an organism's interactions with representations which are in a causal relation with the changes in the environment), then linguistic signs appear to be not signs of components of the medium (that is, physical entities or objects) but of representations which themselves are signs by definition. Consequently, *language is a system constituted by signs of signs.* In turn, linguistic signs as a totality of environmental components constitutive of an organism's particular (communicative) behavior as a description of the niche of cognitive interactions, are subject to change as long as the organism and the medium are in a state of reciprocal causality. This explains, in particular, the historical (evolutionary) nature of linguistic change. In addition, since language is a specific medium (a subdomain of the cognitive domain of interactions), or a part of the world, and at the same time an activity of which this medium is the product, its state is determined in a quite natural way by the state (the sum total of specific features) of the world, which helps understand the developmental history of human languages and their diversity.

The explanatory potential of autopoiesis applied to the core issues in linguistics, semiotics and biology (with other areas of research by no means excluded) is yet to be appreciated, and it is going to be some time before its criticisms give way to a reanalysis of basic theoretical assumptions on which humanistic science has based its exploration of humans and humanness. It is going to be a hard and lengthy process because, according to Boden (2000: 141), to take the philosophy of autopoiesis seriously "would be to undermine many concepts and theories familiar within cognitive science". Yet such undermining is inevitable in the evolutionary development of science (Kuhn 1962) and, as I have tried to show, it has been going on for some time already, particularly, in second generation enactive cognitive science. What is important, however, that this undermining is not in itself an altogether destructive process, it is accompanied by the erection of a new scaffolding with the help of which solid bridges and strong causeways can be built across the gaps and marshes that have heretofore separated linguistics, semiotics, and biology.

CHAPTER 3.

AUTOPOIESIS AND LINGUISTIC ANALYSIS

3.1. A New Epistemology

In the cognitive paradigm, linguistics as the study of language strives to answer the question about the nature of the relationship between language and mind (brain). However, viewed from this perspective, the cognitive enterprise remains largely disorganized in the sense that a synergistic integration of the various disciplines that examine the mind and brain and the role of language as a cognitive capacity of humans has not been achieved, and the "true science of language" has not emerged. That is why the suggestion made by Walenski and Ullman (2005: 328) seems to be well justified:

> Language should be investigated not so much as an interdisciplinary exercise, but rather as a single enterprise that incorporates theoretical and methodological approaches both *across and within* disciplines, including neuroscience, psychology, and theoretical linguistics.

At the same time, caution should be taken when bringing together theoretical approaches in neuroscience and psychology on the one hand, and linguistics on the other; while the intrinsically biological nature of the former two has always been taken for granted, the latter, as represented by major schools of thought over the 20th century, appears to have had very little in common with biological science — except, of course, the generativist claim that language is a 'mental organ'. True, what generativists say about language faculty as an evolutionarily conditioned genetic endowment of humans may, after all, be not such a wild shot. Still, one cannot help being skeptical about their claims — which Paul Postal (2004) describes as 'wide off the mark' — that generative grammar studies biology on the grounds that language is a 'mental organ'. As Daniel Everett (2005:160 ff) sarcastically observes in his critique of *The Language Organ* (TLO),

> [I]f Anderson & Lightfoot are correct, Chomsky deserves the Nobel Prize in Medicine, for single-handedly re-opening the apparently moribund science of anatomy and discovering an organ of the human body which has been overlooked in the history of studies of the human body. Of course, the language organ is unusual in that it cannot actually be seen. What could they mean, you ask, by an invisible organ?
> Surprisingly, the nature of this 'language organ' never emerges clearly in TLO.

But this is only to be expected if the 'biological perspective' on language, advocated for by Anderson and Lightfoot, doesn't go beyond the anatomical and physiological make-up of humans. Looking for 'mental organs' inside the brain which itself is an organ integrated in the nervous system whose function is to control the different aspects of the functioning of the living organism, is hardly a gratifying endeavor. In regard to language as human species-specific behavior ('communication'), taking a 'biological perspective' is tantamount to viewing it as a kind of cognitive activity with *a distinct biological function.* Here the focus should be not so much on the narrow notion of function as specific activity attributed to a particular organ as in the anatomical analysis of the body of an organism, but rather on the broad idea of function as a *specifying feature of the living organism as a whole*. In autopoiesis (as outlined in the previous chapter), living systems are unities of interactions which exist in an environment, they are characterized by circular organization, and this circular organization produces a homeostatic system whose function is to produce and sustain the circularity of organization. But the realization of this function depends on the *components* which determine it and the continuous production of which is sustained by the circular organization. All the particular features of various kinds of organisms *overlay and support this fundamental circularity by sustaining its uninterrupted continuity in sequences of interactions with the constantly changing medium.* So, if we really want to take a biological stance toward language, we should ask ourselves the question, "What is the role of language in sustaining the uninterrupted continuity of a human organism in its sequences of interactions with the constantly changing medium?"

In biology, the more complex the level at which one seeks to explain a living system, the greater the need to examine the network of interactions that lie behind the genome (Cornish-Bowden and Cárdenas 2001). As emphasized by Cornish-Bowden et al. (2004: 716), "the fact that a complex network of interactions connect genes to phenotypes emphasizes the idea that only through the understanding of the whole can we understand the function of the parts". Yet in theoretical linguistics such as structuralism or generativism there is a profound lack of understanding language as a whole, and the question "What is language for?" does not seem to be a priority.

As has been mentioned in Chapter 2, Charles Morris was among the first to suggest that, for languaging humans, 'the response to things through the intermediacy of signs is biologically a continuation of the same process in which the distance senses have taken precedence over the contact senses in the control of conduct in higher animal forms'. Thus he explicitly addressed the biological function of language as one of 'control of conduct' when

> the process of taking account of a constantly more remote environment is simply continued in the complex processes of semiosis made possible by language, the ob-

ject taken account of no longer needing to be perceptually present" (Morris 1945: 32).

With this in mind, the central maxim in the biology of cognition, "Everything said is said by an observer to another observer", appears to be an epistemological recapitulation of Morris's insight on the function of language as a semiotic system.

As was argued in the previous chapter, the advances made by contemporary science in the study of semiosis have been facilitated by an understanding that life is, essentially, a semiotic process. Jordan Zlatev's (2003) unified theory of meaning defines meaning as "the relationship between an *organism* and its *environment*, determined by the *value* which particular environmental *aspects* (falling into *categories*) hold for that organism". The categories into which these environmental aspects fall, are physical aspects, perceived via *innate value systems*, and cultural aspects evaluated on the basis of *conventional value systems* consisting (predominantly) of *signs*. Both innate and acquired value systems serve as *control systems* by directing and evaluating the organism's behavior and its adaptation. Without taking into account the reality of these two systems which, in the case of languaging animals (humans), are integrated on the basis of the relationship of reciprocal causality between them as a result of a history of fine ontogenetic structural couplings between an organism and the environment, that is, without taking an *integrationist approach* to language as a biological phenomenon which supports the 'fundamental circularity in the organization of the living system by sustaining its uninterrupted continuity in sequences of interactions with the constantly changing medium' — hopes for the 'true science of language' to finally emerge will continue to remain just that — hopes.

Building on this assumption, *integrational linguistics* and the *distributed cognition* frameworks have been trying to work out a conceptually more promising program for the study of language and cognition, raising issues of mutual interest and looking for common theoretical ground. As a result, a new approach to language and cognition has emerged, a *distributed language* framework (see Chapter 2). At the same time, much of what the distributed language view offers as interesting theoretical insights on the nature of language and cognition, bears conceptually and terminologically on the basic tenets of autopoiesis as an *alternative epistemology* (Kravchenko 2001a). This new epistemology departs from Cartesian rationalism with its 'idealization' of the mind in that it understands and interprets cognition as a biological process (Maturana and Varela 1980; 1987; Varela, Thompson and Rosch 1991; Järvilehto 1998; 2003a; 2006; Letelier et al. 2003): a theory about living organization is at the same time a theory about cognition (Heschl 1990). Several approaches to language and cognition discussed in the previous chapter, which are embraced by the term *enactive cognitive science* because they all share the basic idea that man and his cognitive abilities should be studied as a biological phenomenon, appear to have subscribed to the biological concept of cognition without necessarily making reference to autopoiesis as their common epistemological framework.

In what follows I will offer a short discussion of some essential properties of language in the framework of the bio-cognitive philosophy of language, hoping to show that this approach may serve as a methodological consensus both for distributed cognition and integrational linguistics, as they all share some central notions and ideas - as becomes obvious in the distributed language approach. At the same time, this approach leaves hardly any room for the code model in understanding language and communication.

3.2. Representations and Signs Revisited

3.2.1. Representations

The conceptual framework of traditional linguistics, much influenced by the 20th century logicians and first generation cognitive scientists, is determined in a decisive way by the assumption that linguistic signs represent different aspects of reality. According to this orthodox doctrine, as Cowley and Love (2006) emphasize in their critique, specialized mental capacities are harnessed to acquiring and storing mental representations of linguistic objects which, by uniting a form with a meaning, allow the transmission of thoughts from one mind to another. As argued by Lamb (1999: 9), beliefs that "our minds contain words that we use when we speak, along with rules for combining them", are nothing but wildly held illusions. Unfortunately, some illusions are hard to dispel, and orthodox linguistics continues to cling to the dualistic concept of representation used in analytic philosophy which has shown its utter incapability to provide a reasonably consistent and empirically sound unified theory of mental representations.

As an example, the orthodox linguist sees representation as the basis for understanding in human communication, for language expresses messages which are "mental representations in the form of conceptual structure" (Pinker and Jackendoff 2005: 205). By contrast, autopoiesis uses a *radically different* concept of representation (see Chapter 2); they are viewed as *relative neuronal activities characterizing the state of an organism's nervous system as a structure-determined system.*[7] Such understanding of representation reconciles the two combative views

[7] A very similar understanding of representation was later proposed by Derek Bickerton (1990: 76): "Representation = 'responding or having a permanent propensity to respond to *x*, an entity or event in the external world, in terms of *y*, a particular pattern of neural activity'". For Walter Freeman (1995), representations are constructed on the basis of the constantly emerging and developing self-organization of neuroactivity, they are "actions into the world", not the contents of an embodied mind. Likewise, Stephen Cowley (2004) stresses the dependence of human represen-

on representation held by integrational linguists and distributed cognition theorists as discussed by John Sutton (2004).

Developmentally, the rise and function of representations as relative neuronal activity may be hypothesized as follows. In the case of a young human organism (a baby), experienced perceived events result in neuronal activity triggered by different types of sensory stimuli co-occurring in real time. With time, an instance of such neuronal activity becomes an 'elementary' representation (ER) in the sense that every time an organism is exposed to a similar 'individual' external cue, it is 'recognized' as something which stands in a 'causal' relation to the ER which is related to another similarly caused ER, and yet to another one, and so forth. When a particular combination of similar stimuli happens recursively over a particular period of time, it results in elementary representations forming a complex representation (CR), which, basically, reflects intuitively inferred 'causal' relations between/among the stimuli recursively[8] co-occurring in real time.

For example, every time during breast-feeding a baby sees the caretaker making a certain gesture (which results in an ER in the baby's nervous system), feels her touch (another ER) to which the baby gives a motor response (yet another ER), perceives a certain smell (another ER) and taste (another ER), and hears at the same time a certain sequence of noises an adult person would identify as a 'word' or a 'phrase' (another ER) — all these (plus elementary representations caused by the baby's bodily inner states), eventually, become integrated into one complex representation characterized by a certain degree of relative stability over time. (Importantly, none of these elementary representations originally takes precedence or acquires unique significance over the others; particularly, as Vygotsky (1987) noted, initially words do not differ in principle from other stimuli given in experience or from other objects with which they become associated.) As a result, even in the absence of other stimuli, one stimulus (visual, auditory non-linguistic, auditory linguistic, olfactory, gustatory, tactile) often turns out to be enough to activate a corresponding CR which incorporates elementary representations of interactions with the absent stimuli. In other words, aspects of the world (including so-called 'words') are sensory stimuli that shape the neural environment, and (complex) representations are run through with motor patterns and familiar co-variants.

tational capacities on environmentally driven neural synergies. However, the traditional 'telementation' doctrine continues to hold strong in much of the literature devoted to the nature and function of language.

[8] Of course, there is no real 'recursion' of one and the same (identical) stimulus over time — rather, we should speak of a more or less adequate match between the already existing ER and the recurring stimuli which are recognized as co-variants of the stimulus that originally triggered a particular neuronal activity.

3.2.2. Phenomenology of Signs

As an obvious consequence arising from the autopoietic conception of mental representations,[9] and much in the spirit of the distributed language approach, the signifying function of linguistic signs does not arise from their direct relation to the external world, it arises from *human experience* as the basis of knowledge. As observed by Love (2004: 531) in his comments on the integrationist's view of linguistic signs,

> [T]he linguistic sign, whether spoken, written or manifest in any other medium, is not an object, or a permanent property of an object. It has no fixed or determinate semiotic value. It becomes a sign as and when used as such, and its signification is a function of that use.

Building on the integrationist work (Harris 1996; Thibault 2000; Love 2004), Cowley (2004) and others take the view that language cannot be context-free and that every contextualization is unique. For Roy Harris, this is the integrational view of 'semiosis', and for Stephen Cowley, an empirical fact. Autopoiesis, however, makes this *necessarily* so. As I have argued elsewhere (Kravchenko 2003b), linguistic signs *cannot be arbitrary* in the accepted sense of the word. The essential non-arbitrariness of signs is sustained by Peirce's semiotics, particularly, by his concept of indexical signs. Peirce used the term *index* for

> a sign, or representation, which refers to its object not so much because of any similarity or analogy with it, nor because it is associated with general characters which that object happens to possess, as because it is in dynamical (including spatial) connection both with the individual object, on the one hand, and with the senses or memory of the person for whom it serves as a sign, on the other hand (Peirce 1932: 170).

As may be seen, the emphasis in this definition is laid on the *phenomenological nature of signs* (including linguistic signs), that is, their cognitive dynamics - the one aspect that until recently has been outside the scope of interest of language sciences. However, even Heraclitus paid attention to the trivial empirical fact about the world (and, by inclusion, about language as part of the world) determined by the phenomenological basis of cognition, in his famous observation, *Everything is in a flux*.

The study of indexicality as a property of signs at different functional levels in the framework of a unified theory of indication (Kravchenko 1992) has shown

---

[9] The fact that the term *representation* has at least three distinctly different readings — in Peirce's semiotics, analytic philosophy, and autopoiesis — may, at times, be confusing, so I will specify it where necessary as'semiotic', 'philosophic', and 'autopoietic'.

that this property is characteristic of *any* sign, albeit to a different degree: thus, there are no pure symbols as Peirce stressed at the time. As Hilary Putnam (1975/1986: 234) observes,

> indexicality extends beyond the *obviously* indexical words and morphemes (e.g., the tenses of verbs). ...Words like 'water' have an unnoticed indexical component: 'water' is stuff that bears a certain similarity relation to the water *around here*.

Bert Hodges (2007) quotes Verbrugge (1985) who claimed that nouns are better understood as indexical, rather than symbolic: "Linguistic actions are not arbitrarily related to their settings or to other actions of the person who is talking. The relationships are systematic and constrained; they are properties of social events that are perceptible" (p. 179).

Admitting the indexical nature of (linguistic) signs is tantamount to questioning their inherent symbolism as the one and only decisive factor in understanding how language works and how it is related to the mind - which, for example, has been the trade mark of first generation cognitive science in general, and generative linguistics in particular, where intelligent performance is viewed as certain symbolic processes involving representations (in the philosophic sense) (Fodor 1975; Newell 1990; Pylyshyn 1999). At the same time, it highlights the external grounding of signs in the physical world and its role in building computational models which, as Angelo Cangelosi (in press) argues, provide an integrative vision of language where the linguistic abilities of cognitive agents are strictly dependent on other social, sensorimotor, neural and cognitive capabilities. The external grounding of linguistic signs allows for novel approaches in understanding the cognitive dynamics of linguistic activity (communication) and alternative views of 'mentalist' (*a la* Jackendoff 2002) accounts of its different aspects. For example, Linell's (2007) concept of 'potentiality' within the dialogistic approach which stresses relational processes in the individual's interaction with others and in general with the environment, bears directly on inherent indeterminacy of linguistic signs as it admits their partial openness and interdependence with contexts: rather than 'encoding' their meanings, linguistic expressions index, cue or prompt understandings in terms of reference, conceptualization and intervention. This view is similar to the idea expressed by David Rumelhart (1979) and supported by Jeffrey Elman (2004) that words don't *have* meaning as much as they provide *cues* to meaning, they must be treated as stimuli whose 'meaning' lies in the causal effects they have on mental states. Interestingly, 'cuing' is one of the central notions in the bio-cultural theory of meaning which distinguishes between four types of meaning systems, *cuebased, associational, mimetic* and *symbolic*, forming an evolutionary and epigenetic hierarchy (see Chapter 2).

Positing phenomenological nature of meaning and, as a result, of (linguistic) signs, goes counter to the concept of automatic (machine) processing of language which first generation cognitive scientists related to syntactic operations on abstract symbols. Elementary representations (relative states of neuronal activity) of interactions with signs, becoming integrated into the network of complex representations, form a system which, essentially, is a cortical network, so language *uses the same cortical structures and processes as other cognitive skills* (Lamb 2006). It would be odd to describe, for example, such adaptively adequate sensorimotor activity as cross-country running in terms of computations on abstract symbols representing (encoding) all the relevant aspects of the physical environment in which the running occurs. Yet this is precisely what orthodox linguists do when they describe language processing — just because to them, linguistic signs are abstract symbols with fixed meanings. But are they, really?

3.3. Interpretation as Meaning-Making

3.3.1. Sign and Meaning, or What's Amiss with the Semantic Triangle

Objectivism. The idea that signs, if their presumed purpose is to serve as a socially approved stable means of storing information in the form of conceptual structure and transferring it through communication, must of necessity be abstract entities, is very appealing because it is supported by the conceptual metaphor COMMUNICATION IS TRANSFER OF IDEAS built in our everyday language — and linguists, just like other people, also 'live by metaphors' (Lakoff and Johnson 1980). To many, most linguistic signs identified as symbols in the spirit of Ferdinand de Saussure, i.e. as arbitrary signs with conventional meanings, indeed seem to meet this condition, and this circumstance did, in fact, facilitate the unprecedented spread and dominance of generative grammar in its different versions as an aspiring rationalist enterprise that marked the emergence of first generation cognitive science (Gardner 1985).

Schematically, the conventionality of meaning is illustrated by the famous semantic triangle (Ogden and Richards 1923/1927: 11) in Figure 1. As suggested by the diagram, between the Symbol and the Thought and between the Thought and the Referent there is a relevant relation, whereas between the Symbol and the Referent there is no relevant relation other than the indirect one, which consists in its being used by someone to stand for a referent (or to 'denote'). If language is viewed as an autonomous sign system (which is characteristic of objectivism), such understanding of linguistic signs does not cause any particular objections. Yet if we reject Saussure's principle of distinguishing between *langue* and *parole* as to a large degree artificial and empirically unsustainable, certain doubts arise about the general empirical validity of the semantic triangle.

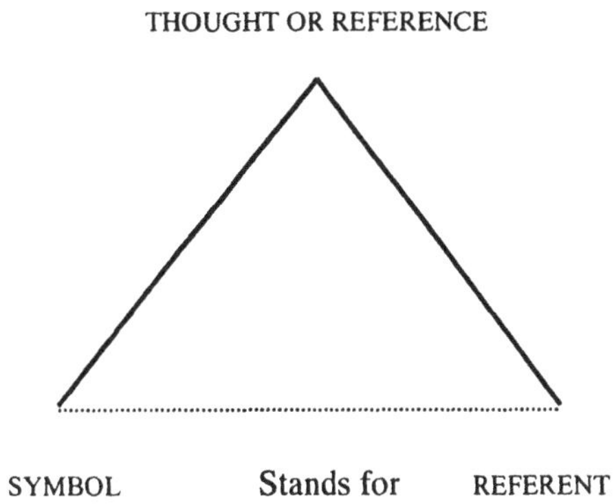

Figure 1. The semantic triangle

First, the definition of linguistic sign as a unit of language used to denote things and phenomena in the world does not answer the question in what relation language itself stands to these things and phenomena. Ontologically, language seems to be both a dynamic phenomenon (communicative activity) and a thing, that is, a system of 'linguistic items' identified and singled out by the classical linguist as a result of his efforts to 'construe' language from texts (which are a kind of thing, after all) that can be described as sequences of conventional graphic symbols 'organized' according to the 'rules' of a particular language and believed to be representations or translations of acoustic phenomena associated with natural language, into stable static forms typically viewed as a code. However, this conceptualization of language as a thing (a code) 'out there' is one of the main obstacles to understanding important matters concerning language, mind and linguistic behavior (Cowley and Love 2006). As Don Ross (in press) argues, language is not a set of structures from which one can recover, by de-coding, 'ultimate' information grounded in some more fundamental format. Besides, although language as a specific kind of human bio-socio-cultural activity may, probably, be described as *epiphenomenal* (that is, as a trivial and accidental byproduct of truly significant processes, such as the biological evolution of the human species from hominids), to every single member of a given community speaking a given language linguistic phenomena appear as *natural components of the medium* (including both the physical and the cultural) of which this community itself is a part. To say that the function of linguistic signs is to denote things and phenomena in the real world is akin to saying that the function of a beard or a moustache is to denote a male human, or that the function of tears is to express sorrow.

Second, words as typical linguistic signs (natural acoustic phenomena as well as their graphic representations) possess physical substance at every single moment of their existence. They stand in spatio-temporal relations to various physical parameters that shape the context of every word's being, or its environmental medium. It is, therefore, quite obvious that changes in the physical context result in changes in the nature and number of a particular word's relationships with other phenomena constitutive of its environment. It follows that the set of associations responsible for the emergence of what can be called 'experience of sign' and, because of that, bearing directly on the sign's 'content', may vary.

Third, the being of sign as a physical entity in space and time implies its physical change: the form of the sign is not permanent, given once and for all, it is subject to considerable permutations — which has been more than convincingly shown by historical linguistics. These changes, by return, cause qualitative and quantitative changes in the relationships between sign and other entities including other signs, and thus cannot but tell on the resulting experience of signs which the sign user relies on, and on the 'conceptual structure' which constitutes, so to speak, the 'content' of the sign form. As a consequence, the relationship between the sign vehicle (the signifier) and its content (the signified) at every given moment cannot be characterized as absolutely arbitrary insofar as we cannot describe as arbitrary the nature of associations which accompany the sign and are determined by the user's experience of the sign; therefore, *it is not a relation of coded equivalence*.

Fourth, these associations may be described as *concepts* or *conceptual structures* prototypically understood as mental structures functioning as operational units of consciousness in the process of acquiring and representing (in the autopoietic sense) experience. Therefore, the physical nature of signs on the one hand, and the phenomenological roots of concepts represented (in the semiotic sense) by signs, on the other, do not allow for treating signs as pure abstractions (cf. Radden and Panther 2004). But if it is indeed the case, then the meaning of any linguistic sign cannot be described completely and exhaustively, and the concept of representation (in both the semiotic and philosophic sense) as the function of language in general and linguistic signs in particular, is simply inapplicable. As noted by Chris Sinha (1999: 228 ff), the problem of representation "simply *does not arise* if we cease to view meanings as immaterial objects".

Another strong argument against semiotic objectivism bears on the relationship between the adjoining sides of the semantic triangle which appear to be equipollent in Figure 1. If the relationship between the sign and its physical referent in reality[10] is mediated by thought, and if thought, in general, is understood as intentionally driven processes in the human brain, while mind is seen as the totality of states of specific neuronal activity at every given moment both determining, and determined by, such processes (whether we call them mental, neuronal, electro-biochemical or whatever), then we inevitably face the question of *directionality* of the vector of intentionality for each of the sides of the semantic triangle, particularly, for the sides which, according to Ogden and Richards, 'represent' the relevant relations between THOUGHT and REFERENT and between THOUGHT and SYMBOL. Importantly, intentionality in this case is not the ill-famed 'aboutness'; in keeping with the epistemological principles of autopoiesis, intentionality is described as *the capacity of a living system to modify its state of reciprocal causality with the world on the basis of experience acquired with time for the purpose of sustaining the ecological system which enables reciprocal causality between the organism and the world* (we will take a closer look at intentionality thus understood in Section 3.4).

What is the issue of directionality about? If every sign ('symbol', in Ogden and Richard's notation) is a manifest entity which stands for another, non-manifest entity because of the relationship between the two that emerges experientially in the course of cognitive interactions with the world and is rooted in consciousness as defined in the previous paragraph — that is, if both the sign and that which it stands for are physical entities in the same time-space and, because of that, perceptually accessible to the cognizing agent, what then is the directionality of the vector of intentionality connecting the sign and the referent? For objectivism, the conceptualizing activity of a conscious agent is directed from the referent to the thought about this referent, and from the thought ('image', 'mental representation') to the sign. By standing in for the referent, that is, by 'representing' it, the sign, in its turn, makes it possible for the thought to 'refer' ('to be about', in the objectivist philosophy of mind) — hence the concept of signs as tools called upon to serve the thought as the representational feature of consciousness. What this means is that, with the objectivist approach to the problem of representation, the semantic triangle should look as shown in Figure 2.

However, such modification of the sign relationship diagram does not remove the problem of defining the representational function of sign as it does not take into account *agentivity,* or the intersubjective nature of semiosis on the one hand, and ontological properties of signs and their referents, on the other (Kravchenko 2003b). For a component of the environment (a potential referent of a potential sign) to be taken into account by a cognizing mind, that is, for it to be *intentionally* brought into an organism's domain of interactions with the environment, its

[10] It must be noted that 'reality', being the core concept in objectivist philosophy, appears to be a relative notion in the epistemology of autopoiesis with its foundation in Maturana's theory of perception and structural determinism (Imoto 2004a, 2004b, 2005).

mere spatio-temporal existence in the domain of immediate experience is not sufficient — it must be 'taken into account' by the organism as a *causal factor* which affects sustainability of the ecological system "organism-niche". As it is, not all environmental components constitutive of the niche acquire the 'causal factor' status in various situations of interactions of an organism with its niche. For it to happen, an environmental component must become *salient* against the background of other components constitutive of the organism's niche, it must possess *value* which is by no means absolute but *relative* and depends on specific situational characteristics. Consequently, in the case of the semantic triangle the side which 'represents' the relationship between THOUGHT and REFERENT must show its bidirectionality. However, as soon as we make this rather trivial inference, we cannot do other than acknowledge the decisive role of *consciousness* in construing any physical entity as a referent, particularly, the referent of thought. And this leaves us with no choice but reject the objectivist interpretation of the sign relation which, in this case, becomes impossible by default.

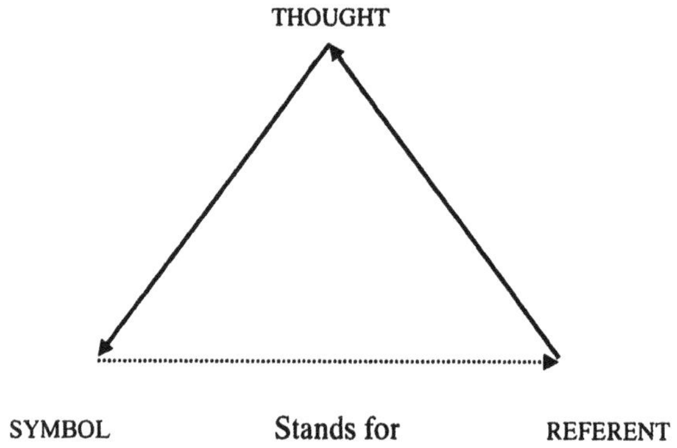

Fugure 2. Vectors of intentionality in the objectivist interpretation of the semantic triangle

It should be mentioned that in the first quarter of the twentieth century, Ludwig Wittgenstein (1921/1961) attempted to solve the "problem of representation" born by the dogma of autonomy of linguistic meaning (that is, the problem of how meanings as objects existing in the sphere of the Ideal, reflect or represent objects in the physical world), and came to a conclusion that the problem was insoluble. As argued by Chris Sinha (1999: 228), the "problem of representation" cannot be solved in a satisfactory way as it "ties together the Two Dogmas of Reificatory Semantics into a philosophical Gordian knot" (see Chapter 1).

Subjectivism. Opposite to the standard objectivist theory of meaning is an approach whereby things referred to by linguistic signs are viewed as created by language itself, that is, sign and thought are taken to form a unity, and thought and operations on linguistic signs form a complex twofold process (Shaumyan 2001). Such view of language was propounded by Wilhelm von Humboldt (1820) and received further impetus in the writings of Edward Sapir (1929) and Benjamin Lee Whorf (1956). The extreme manifestation of this approach is *subjectivism* or the belief that meanings are 'in the head' in the form of 'mental objects' and that the only reality is the reality of what is known under the name "mind/brain". Close to this approach is the concept of *embodied mind/language* (Johnson 1987; Damasio 1999; Lakoff and Johnson 1999 *inter alia*) and the *neural theory of language* (Feldman 1982; Ajjanagadde and Shastri 1991; Regier 1995; Narayanan 1997 *inter alia*).

The subjective approach to the problem of representation can also be illustrated by a modified semantic triangle. For the human mind to reflect a thing from the physical world as an object of thought (referent), it must single out this thing from the universe as a discrete entity. Once it has been done, the THING — THOUGHT relationship takes a different form, as may be seen in Figure 3. Vector A_1 symbolizes the initial impact of an environmental component (THING) on the organism, and Vector A symbolizes its intentional categorization as a causally relevant (meaningful) environmental factor which must be taken into account in the organism's current (but not necessarily all the subsequent) interactions with it. This meaningfulness is 'represented' (in the autopoietic sense) by a specific state of neuronal activity in a specific area of the cortex, which acquires a semiotic status by becoming a *sign* (more on this in the following subsections).

At the same time, subjectivism assumes 'producibleness' of linguistic signs in the sense that they are treated (at least, implicitly) as the products of the mental-linguistic activity (Vector B) specially produced for the carrying out of the symbolic function, that is, the 'transfer' of meanings. However, this assumption is wide off the mark. For every single language user, what we call 'linguistic signs' is simply one out of many kinds of *natural* physical entities (things, phenomena, etc.). As argued by Donald Davidson (2001), the meaning of words depends on the causal history behind the learning of these words. By interacting with words as a kind of physical entities, we discover their meanings as certain significant relationships with other manifestations of the physical world (such as THING_1 and THING_2 in Figure 3), including *homo loquens* in all the aspects of his agency, such as mental activity (THOUGHT), for example, and first and third person authority. Yet the *other mind factor* (THOUGHT_2) finds no room in the classical semantic triangle, and without it any schematization of semiosis can be only incomplete or/and distorted.

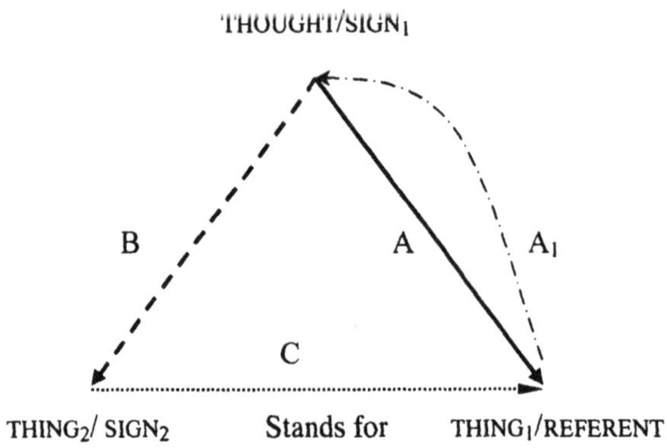

Figure 3. A subjectivist model for the sign relation

Both the objectivist and subjectivist approaches to the problem of representation proceed in their construals from language understood as a *sign system*, while any and every meaning emerges only in the process of humans' interactions with the environment, its nature lies in human activity determined by the very being and adaptive behavior of humans as a *social species*. We should, therefore, speak of an *intersubjective approach* that would take into account the multiple minds which, after all, make semiosis possible. Besides, for every individual human cognizer any linguistic sign is, ontogenetically, a component of the (ecological) niche occupied by them, it is an environmental constituent, and the dynamics of interactions with linguistic signs do not differ in principle from the dynamics of interactions with other kinds of physical entities which can function as referents of signs. Every time a human cognizer interacts with a natural linguistic sign as a physical environmental component which becomes, as a result of such interaction, part of the meaningful relationship THING — THOUGHT, he interacts, one way or the other, with the *source* of the sign as an acoustic phenomenon, that is, with another organism and another mind. It means that the meaningful relationship THOUGHT — SIGN emerges as a consequence of establishing an initial meaningful relationship between two organisms (minds), the latter relationship being *biologically* determined (see Chapter 2 section 5.3).

Intersubjectivity. The interpretation of linguistic signs as objects, characteristic of the traditional linguistic paradigm, is nothing but a *deep delusion*. It is because of this delusion that both objectivist and subjectivist semiotic construals turn out to be absolutely untenable. The understanding of language as a species-specific biological activity accompanying social behavior calls for an understanding of linguistic meaning as rooted in this very activity. The schemas presented above in Figures 2 and 3 are based either on the assumption that signs as a kind of physical entities are given *a priori* (objectivism), or that signs are the products of intentional processes in the mind (subjectivism). As it is, the truth, as often is the case, lies in between. As a manifest physical entity, any linguistic sign is 'present' to the mind via perception, but it is present (or 'given') precisely as a component of the physical environment which may or may not become a causally relevant factor in the current situation of an organism's interactions in which it is involved. While giving a component of the physical environment (an epiphenomenal acoustic phenomenon) the status of causal factor does not change its physical nature, it draws it into a *dynamic system of relationships* between the two interacting organisms (minds). These relationships are *orientational relationships* inasmuch as interactions between the organisms aim, in the long run, at sustaining the ecological (reciprocally causal) system "organism-world".

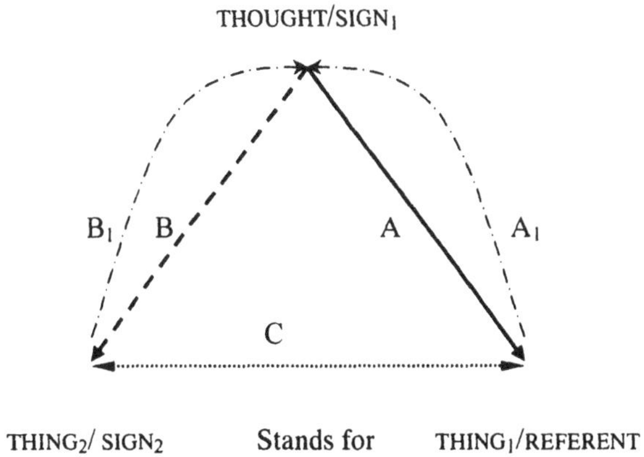

Figure 4. An intersubjectivist model for the sign relation

As is shown in Figure 4, and similar to the relationship A — A$_1$ (THING$_1$/REFERENT — THOUGHT/SIGN$_1$), neither the relationship B — B$_1$ (THOUGHT/SIGN$_1$ — THING$_2$/SIGN$_2$), nor the relationship C (THING$_2$/SIGN$_2$ — THING$_1$/REFERENT) are unidirectional. Moreover, the semantic triangle gives a truncated picture of semiotic relations as it does not take into account other organisms as a prerequisite for the genesis of linguistic meaning. But this is something only to be expected, since a triangle is a rigid geometrical figure which does not allow for a fourth vertex.

Acknowledging intersubjectivity of linguistic meaning which, for Chris Sinha (1999), emerges in the "shared universe of discourse <...> continuous with (not separate from) the material world in which other (non-discursive) human activities are carried out" (p. 232), leads to a natural inference about the *multifariousness* of meaning as a cognitive phenomenon arising in the cognitive domain of interactions of organisms, or in a *consensual* domain. Language acquisition is a biologically grounded socially determined process whereby words, sentences and the structure of the situation (physical context) become objects of *simultaneous perceptual experience* (Strømnes 2006). It is for this reason that the assumptions of propositional logic are inapplicable to natural language or, to be more precise, when applied, they turn out to be erroneous. It is time to reject the view of communication as an activity that consists in the transfer of 'meanings' (mental representations) and to view it instead as efforts to establish a system of shared *points-of-reference* in an interactional situation (consensual domain).

3.2. Signs in Physical Context

Characteristic of the mentalist approach, which is intrinsically denotational, is the relation of the meaning of a sign to its capacity to activate in the interpreter's mind a concept, or a set of concepts (conceptual structure) associated with this sign (S. Gross 2005). On a mentalist account found, for example, in Aitchison (1987), words we hear ('physical' words) are matched against words stored in our minds, and native speakers carry out a search of their word-store when they need to recognize a real word or reject a non-word: "The large number of words known by humans and the speed with which they can be located point to the existence of a highly organized mental lexicon" (Aitchison 1987: 9). Thus, two sign systems are posited, one 'out there' in the physical world, perceptually shared by members of the community, and the other in the head of every individual language user. The latter is a (mental) reflection of the former, so it appears that while the size of an individual word-store may vary from speaker to speaker, the meaning of each 'mental' word stored in the head of one speaker is not different from the meaning of such stored in the heads of other speakers, and this fact accounts for understanding others. However, such view is untenable for the following reason.

It is an empirically trivial observation that for whatever is called 'concept' to be associated in the mind of the hearer with the perceived sign, the actual decision as to which concept out of the great number of concepts allegedly 'stored' in the so-called 'mental lexicon' it should be, must be made on *the same grounds* as when the sign user makes a choice in favor of one out of a number of signs potentially available for the representation (in the semiotic sense) of the same concept — otherwise, the resulting relationship between the sign and the concept it stands for (denotes) for the hearer will not be one of identity and/or equivalence necessary for understanding the speaker. This understanding (that is, sufficiently adequate interpretation) of linguistic signs is, essentially, a procedure for a more or less approximated identification of such grounds which to a large degree relies on taking into account a large number of variables that are not constitutive of conceptual structure in the mentalist accounts of language, but form the physical context of communication as bio-cognitive activity (activity in a consensual domain). The less exactly these grounds have been identified, the more probable it is that the 'input' knowledge and the 'output' knowledge (to use another metaphor deeply ingrained in the metalanguage of cognitive science) will differ - hence the line drawn between the 'meaning' of a sign and its 'sense' in a particular communicative act.

This realization led to introduction of the term 'speaker's meaning' into metalinguistic discourse because, apparently, there are no two speakers using one and the same identical language system that could be described as a code (see Love 2004 for discussion). Speaker's meaning has been routinely contrasted with conventional meaning as the one deviating from certain norms and rules for sign interpretation accepted in a given linguistic community. But norms and rules are only a reflection of some common generalized experiences of language by the linguistically minded users, they do not necessarily bear on *real facts* about language, they are not imperative prescriptions passed down by someone which must be followed to the letter (although it is very often the case when quite the opposite is true, as with the double negation rule, for example). As Bert Hodges (in press) argues, our utterances and understandings are not fundamentally rule-governed, but values-realizing. As an illustration, we know that beer is a light alcoholic beverage made of water, malt, and hops, and should we observe someone sprinkling her head with beer from an open bottle, it will not for that reason stop being to us what it is, beer. It is a different question what goal the user in this case is trying to achieve (different users may have very different goals), but if we assume that the person we observe is not insane, we will, most probably, start looking for a possible reasonable explanation for such unconventional use of beer, trying to relate what we just witnessed to our *experience* and *knowledge* — which, in the case of the older generation at least, will tell us that a few decades ago women used to use beer in such an unconventional way when curling their hair.

Seemingly deviant uses of signs are very similar in this respect, and if we possess sufficient experience of signs, if we have shared experience with the sign user and experience of the world in which we, the sign user, and the sign exist (that is, if we operate in a consensual domain), we can, as a rule, identify, with a certain degree of approximation, the grounds on which the choice has been made in favor of one particular sign rather than any other from all possible signs (cf. Grice's 'conversational implicatures'). In particular, this process underlies such typical natural language phenomena as metonymy and metaphor. Yet this identification is never a relationship of equivalence by default, and the cognitive effort of the interpreter of linguistic signs (the observer of communicative verbal behavior), although reducing the degree of indeterminacy characteristic of the sign use and hence its purported 'meaning', yields results which have already been to a certain degree predetermined by his overall individual and, therefore, unique experience of the relevant physical, biological, social, historical, cultural, etc. parameters bearing on the communicative situation. This experience characterizes the current state of the language user as an organism (either in the stance of speaker or interpreter), this organism being a *structure determined system.*

3.3.3. Ontogenetic Structural Coupling as the Basis for Meaning-Making

In autopoiesis, structural determinism is defined as follows:

> A structure determined system is a system such that all that happens in it or with it arises as a consequence of its structural dynamics, and in which nothing external to it can specify what happens in it, but only triggers a change in its structure determined by its structure (Maturana 2000: 461).

The structural dynamics of a structure determined system is the result of an organism's ontogenetic history of fine structural coupling with the environment. The cognitive domain of linguistic interactions, which cannot be identical from one individual to another inasmuch as one living system[11] cannot be identical to another living system, is constitutive of this dynamic environment. Therefore, even if we subscribe to the highly questionable idea that there is, in fact, a 'mental' lexicon in our heads, an obvious observation is in place: *the content of the mental lexicon of one individual cannot be identical to the content of the mental lexicon of another individual.* But what is this content?

In traditional linguistics, signs are routinely characterized as forms ('expressions') filled with certain content ('meaning'). Hereby content is understood not as what may, and often is, described in philosophy as the grasped essence of a given phenomenon represented (in the semiotic sense) in this form, but rather as the experience of this phenomenon's relationships with other entities or phenomena in the real world (known as signs) perceived as *epistemically causal relationships.* A red sunset in itself does not necessarily speak of windy weather we may find tomorrow: it can be the effect of a gigantic distant forest or steppen fire, unusual paranormal phenomena in the atmosphere, or the like. It is only the accumulated available experience that allows a human interpreter to find a relatively regular connection between the sunset color and a possible change in the weather; and

[11] In the biological theory of cognition, an organism is a living system as long as it is autoreferential and specified by its circular organization. A cell is simply a first order living system, an individual human — a second order, and a human community — a third order living system.

it is this relationship that becomes the content of the sign. As it were, red sunset becomes a sign in the sense that, on account of available experience, it 'embodies' to the observer certain information without being intentionally produced by man.

A sign user who intentionally uses a sign does not extract it from some obscure depository of sign vehicles to which the so-called conventional meanings have been attached (by whom? when?). He uses signs that have been 'acquired' and 'mastered' in the course of gaining individual experiences as an observer of the world in the spatio-temporal, geographical, historical, cultural, etc. medium of which language itself is a part. Various aspects of these diverse experiences are projected onto a combination of psychic (mental) associations (or concepts) whose attachment to a given linguistic sign acquires public significance. That is why for every sign user every sign is 'processed' in the mind of the observer in the totality of its relationships with very different things and events.

The less prerequisite knowledge of the medium in which the sign user functions as *homo loquens,* the less the total amount of implications this sign may cause. For example, a red sunset is not indicative of anything to a little child who lacks relevant experience; that is why a red sunset will not be a sign to this child in any acceptable sense. But just in the same manner, a word spoken by someone will not be a sign to the little child, either, on condition that it is new or unknown:

> Not only a child, but even an adult individual unfamiliar with some special language (such as the languages used by professionals in medicine, botany, chemistry, mathematics, etc.) is, in a way, not unlike a semantic aphasiac. Words whose external makeup corresponds well to the words of his native language to him seem hard to combine into meaningful sequences (Zhinkin 1998: 66).

In other words, what is, by definition, a produced sign for a given linguistic community, does not necessarily function as such for every single member of this community capable of perceiving this sign. To a human individual, a word becomes a sign only after it has become, as a bonding and mediating component, part of a system of stable associations between things and phenomena in the world, these associations forming a specific mental structure (concept) which, in the long run, forms the basis for what is usually referred to as the 'meaning' of a sign. Words mean nothing of themselves, and meanings are not 'out there'. As Linell (in press) argues,

> it is a category mistake to talk about meanings and meaning potentials as either internal ("in the head") or external ("out there"): communicatively interacting members of a community *partially* share the meanings established in and through how people relate to each other and to objects, processes and circumstances around them.

A system of associations between things and phenomena in the world depends in a very straightforward manner on conditions and circumstances under which the sign is 'acquired'. Therefore, to every single individual a set of jointly encountered entities which causes a certain implication, is uniquely individual. As was noted by John Locke (1961), for every human the words he uses express those ideas which he has and which he would like to express by them. As an example, the utterance in a Russian-speaking community of the word *kasha* ('hot cereal') will cause in the mind of one individual an association with cream-of-wheat sweetened with sugar, for another the association will be with rice boiled in milk, for a third one - with buckwheat, barley, millet or oat meal boiled in water, with salt and butter added, and if the interpreter cannot retrieve these associations from the sign addressed to him, one cannot say that the interpreter has received the full meaning intentionally attached to the sign by the sign user. Bertrand Russell (1948: 4) wrote in this connection:

> You can learn by a verbal definition that a pentagon is a plane figure with five sides, but a child does not learn in this way the meaning of everyday words such as "rain", "sun", "dinner", or "bed". These are taught by using the appropriate word emphatically while the child is noticing the object concerned. Consequently the meaning that the child comes to attach to the word is a product of his personal experience, and varies according to his circumstances and his sensorium. A child who frequently experiences a mild drizzle will attach a different idea to the word "rain" from that formed by a child who has only experienced tropical torrents. A short-sighted and a long-sighted child will connect different images with the word "bed".

Thus our understanding of linguistic signs (such as words) is an interpretive process which crucially depends on the degree to which the sign user and the sign interpreter share similar life experiences, and our meaning-making ability as one of the aspects of the complex cognitive dynamics that characterizes us as structure determined systems, is predetermined by our unique histories of ontogenetic structural coupling with the (linguistic) environment and cannot be reduced to the simplistic notion of mapping 'external' signs (spoken, or 'physical' words) against internal signs (the 'mental lexicon') in search of the so-called equivalence.

To sum up what so far has been said about signs and meaning:

1. Denotation can not be a primitive operation as it requires an agreement consensus, so there can be no *pure* symbol-grounding. Representations as states of relative neuronal activity do not depend on encoding/decoding procedures, they depend on resonance effects because they result from *ontogenetic structural coupling*. In that, they are structure determined and mediated by intentionality.

2. Signs are phenomenological, and their phenomenological givenness should be part of the definition of sign. In signifying, a history links indexical

properties with experience: dynamical connections arise between the indexed and a person's senses/memory; that is how *signs contribute to (experience of) flux.* Counter to the denotational view, signs are characterized by *unfinalizability* and *interdependence with context.*

3. Since words of natural (spoken) language are physical, they help constitute a context. When context changes, so do a physical word's connotations — and, by extension, the word itself. This affects the word's relations with the world and a user's experience. Therefore, we use associations in interpreting 'meaning', and words are not denotational (although one, or some, association could have primacy). The link between form/meaning is thus not arbitrary — it is not a notation that codes experience.

4. Signs are involved in shaping up concepts by setting off neuronal activity as representations of interactions with the ever changing environment with which the organism is structurally coupled or 'braided together'; signs, therefore, contribute to the structural dynamics of the organism as a living system. On these grounds, signs cannot be abstractions, but if they are not abstractions, they have no content and the problem of representation (in the philosophic sense) 'simply *does not arise*'.

3.4. Intentionality and Signification

3.4.1. The Notion of Intentionality

Now I want to focus on the notion of intentionality invoked in the previous section with the objective to explain what the term stands for in the biological theory of cognition and what is to be gained by it in trying to understand and explain language.

The term 'intentionality' is used in modern science at least in three separate areas: philosophy, linguistic semiotics, and autopoiesis as a theory of self-organized living systems. I am not going to dwell on the decades-long debate among rationalists about intentionality as a specific psychological category ('aboutness') which connects thoughts with things and mind with the world (Putnam 1988; Haldane 1989; Priest 1991 *inter alia*) for a rather obvious reason: to me and those who reject the code model of language, the Fodorian (Fodor 1987; 1998) problem of intentionality is a dead issue. As for linguistic semiotics, it defines intentionality as a necessary constitutive property of signs: a sign is an artifact intentionally produced for the purpose of performing a communicative function, and that is what makes it intentional. However, speaking about the intended "producibleness" of linguistic signs, orthodox linguists overlook the essential difference between the spoken and written language (Linell 2005). Admittedly, a written word is a *bona fide* artifact, but how justified is inclusion of spoken words in the class of artifacts?[12] Is it just because one out of a number of existing hypotheses about language origin claims man to be its conscious creator? But a hypothesis, no matter how appealing, remains a hypothesis until it has received sufficiently reasonable and convincing proof. As of today, science does not have solid proof that language is the product of man's purposeful activity.

On the contrary, researchers into the origins of language are inclined to think that language is the evolutionary property of man as a biological species (Hurford et al. 1998), it is not a cultural artifact that we learn (Pinker 1995a), it is a complex adaptation for communication (Pinker and Jackendoff 2005). In this case, the issue of artificial origin of linguistic signs (physical words) is stricken from the agenda altogether. Indeed, it would hardly occur to anyone to speak of artificial origins of laughter, crying, or screams of pain, despite the fact that each of these can be considered as a means of communication as they produce orientational effects in a consensual domain of interactions. Here we are reminded of the inadequacy of the commonly accepted definition of the function of language as one of communication (see Chapter 1 for discussion).

Languaging constitutes only a subdomain of communicative interactions which can be both of linguistic and non-linguistic nature. At the same time, by contrast with non-linguistic types of communication, the use of language (that is, oral speech) is always associated with intention as an act of volition or a desire to enter communication as shared (and, therefore, value-laden) activity. Yet the matter is not that simple. On the one hand, laughter or the sounds of crying do not differ from speech in this respect because a man by a similar act of volition (i.e., intentionally) may make himself cry or laugh even when he does not at all feel like crying or laughing, or he can restrain himself from a compelling desire to weep or roll with laughter. On the other hand, a trivial case of linguistic activity is the so-called "non-communicative" speech act when the addressee is either the speaker himself, or an animal without linguistic ability, or an object known to be inanimate, or there may be no addressee as such. The latter case is illustrated by situations when a man totally absorbed in some activity utters words, phrases, or whole sentences often without even being aware of the fact. The only thing which distinguishes the sounds produced by a human in such a case from non-linguistic sounds (sighs, groans, and the like) is the meaning associated with them and established as part of the 'language system', but they often do not have any identifiable *communicative sense*. If we, however, remember that traditionalists consider intentionality in the context of intended *purposefulness* as the property of a sign addressed to a receiver, then the question arises, "Who is the addressee here?" So,

[12] For a relevant discussion see Wheeler (2004).

the notion of intentionality in linguistic semiotics appears to be indeterminate and fuzzy.

In the biological theory of cognition, and distinctly different from linguistic semiotics, intentionality is viewed as *a mode of self-organization of a living system*: interacting with the environment (as it appears to an observer and without reference to the autonomous unity), an organism adds the surplus of *signification*, so that it becomes the environment *for* the organism which is defined in the same movement that gave rise to its identity and that only exists in that mutual definition — the system's world (Varela 1992). *Signification* in the above sense is very similar to the ecological concept of *values* (Hodges 2007) as the global constraints on self-organizing ecosystems, the boundary conditions that, according to Hodges and Baron (1992: 270), "provide not only for the initial conditions for the system but ... also underwrite the system dynamics". A similar understanding of *value* is found in Jordan Zlatev's (2003) unified biocultural theory of meaning, where it is stressed that only living systems have intrinsic value which is a necessary and sufficient condition for possessing the category *meaning*. The theory also emphasizes that a value system must be *intrinsic* to the system in the sense that it serves to preserve the system's *organization.*

So, if we relate the notion of intentionality directly to the semiotic process of sign generation, that is, to conceptualization and categorization of objects as meaningful entities interactions with which produce orientational influences on an organism by *in*-forming it in the world, then we have no alternative but to consider intentionality as *a biological cognitive function* when it is treated as a cognitive phenomenon of the living organism's self-regulation. By and large, the problem of intentionality cannot be separated from the problem of meaning; moreover, it is the key to understanding the genesis of meaning. For if we assume the causal nature of sign relations, we cannot but acknowledge that a causal relation itself (such as the relation between smoke and fire, for example) is already signification brought unto the world by a living (cognitive) system in its movement to *in*-formation in this world through the mutual definition of its identity in the organism's environment created by signification. In the case of humans, much of this in-formation occurs in *communication.*

3.4.2. Communication

Communication integrates activities that structures interaction and, in humans, language. Rejecting the code model of communication and viewing it as integrated activity in real time poses the following questions: "What is integrated?", "How is this possible?", and "What sets off the process?" While code models derive from a cultural tradition that privileges monologue and uses written texts to define linguistic units, in distributed view on language and cognition language is a heterogeneous set of artifacts and practices that, during ontogenesis, become integral to human activity, and communication is "a process by which semiological values are created *sur place* through the integration of human activities" (Harris 2004: 734).[13] In Stephen Cowley's (2004) terms, our contextualizing bodies integrate activity in real-time as we use socially and culturally defined patterns. Language is, on this view, simultaneously enacted and interpreted across the time-scales of biological and cultural events: language is integrated with brain and behavior, linguistic behavior results from the real-time integration of activities, including ones that serve as signs. In this, the integrational and distributive accounts of communication converge with the biocognitive view on communication.

The essential feature of communication in autopoiesis is that one organism produces an orientational influence on another organism as a result of which the behavioral response of the organism to be oriented (the description of its niche) is modified.[14] It must be understood that this influence is orientational mostly because we can say of the first organism that it has *an intention* to modify the other organism's behavior in a certain way — which, however, does not mean that the second organism's behavior will necessarily be modified. But what is the source of this intention which makes communication possible, and of what relevance can modification of the second organism's behavior be to the first organism? Evidently, such modification purports to change the medium of the first organism (its domain of interactions) of which the second organism with its niche is a part. The purpose of this change is to optimize the first organism's interactions with the medium,[15] i.e., in the long run, it is determined by the first existential priority of a living system - survival: "For the organism, the environment is a set of processes and components that have to be recognized and manipulated in order to be capable to survive and reproduce" (Moreno, Merelo, Etxeberria 1992: 66).

Everything occurs in a physical context (consensual domain); when individuals are co-present, each influences the other. Modification of the behavior of an organism to be oriented is possible under the following conditions:

(i) a consensual domain of interactions for both organisms (context),

---

[13] Note how close this integrationist definition of communication is to Varela's definition of intentionality as 'a mode of self-organization of a living system' given above.

[14] Conceptually, autopoietic view on communication is very similar to Ross's (2004) view on communication as signalling moves aiming at reaching equilibria in coordination games with others. In Linell's (2007) dialogical theory *other-orientation* is one of the key concepts, along with *interaction* and *contexts* — all of which reveal striking parallelism with the autopoietic view on language and communication.

[15] In this respect, Ross' (2004) game-theoretic approach to communication is quite compatible with the autopoietic one.

(ii) the organism to be oriented (the observer) interacts with behavioral patterns of communicative interactions of the orienting organism with representations of first-order descriptions (non-communicative behavior), which presupposes

(iii) an intentional act of bringing signification into the consensual domain of interactions both on the part of the orienting organism and the organism to be oriented; it is this which enables self-regulation of an organism as a living system *in*-forming in the environment and making it the world filled with signification.

For example, organism *A* and organism *B* are in a consensual domain of interactions. Organism *A* interacts with the medium, and its behavior which is an outcome of such interactions is a description of these interactions (a description of the niche, or a first-order description). Interactions of organism *A* (which is an observer of its own interactions) with such descriptions result in specific states of neuronal activity or representations (second-order descriptions). Then, organism *A* enters the domain of interactions with representations, and its behavior modified by these interactions to organism *B* will be a third-order description, that is, a description of interactions with representations (communicative behavior).

Organism *B*'s observation of the third-order descriptions is none other than interactions with environmental components which result in representations, but this time in organism *B*'s nervous system. Organism *B*, in its turn, interacting with the medium not unlike it is done by organism *A*, may itself play the role of an observer of its own descriptions. In such a case it finds itself in a situation when interactions with organism *A*'s third-order descriptions cause a representation in the mind of organism *B* whose configuration more or less coincides with the configuration of a representation which already exists as a result of organism *B*'s interactions with its own descriptions. The resulting 'resonance effect' activates the already existing representation interactions with which lead to a modification of organism *B*'s behavior, this modification having been initially caused by organism *A*'s interactions with its own representations.

This is, basically, what communicative interaction is about, and as may be seen, its structure is rather simple. Yet one link in this structural chain remains a mystery — the driving force behind an organism's interactions with representations of its own interactions. What makes an organism enact such behavior?

3.4.3. Reciprocal Causality

The answer evidently lies in understanding that representations as specific states of neuronal activity possess signification in the sense that they modify an organism's behavior bearing on interactions with another organism as a component of the world with which the first organism stands in a relation of *reciprocal causality*.[16] Positing reciprocal causality as an ontological given allows us to speak of "various *modes* of self-organization where the local [an organism] and the global [the world] are braided together" (Varela 1992: 6). The significance of reciprocal causality consists in that a change (as a result of interactions) of one element in the relation effects a change in the other element, modifying the medium which, in turn, exerts a modifying influence on the organism, and so on in recursive order. And here again we are dealing with circular organization:

⇒ interactions with the medium are described by organism *A*'s behavior

⇒ organism *A*'s own observed behavior is described by representations (in the autopoietic sense)

⇒ representations are described by organism *A*'s communicative behavior

⇒ organism *A*'s communicative behavior observed by organism *B* activates representations in organism *B* which are descriptions of its own behavior analogous to organism *A*'s behavior

⇒ organism *B* modifies its behavior as if it were interacting with the (roughly) same medium and in the same manner as organism *A*; but because the domains of interactions of *A* and *B* do not coincide completely, the modified behavior of *B* cannot be a complete description of its domain of interactions, it will only describe the domain of communicative interactions changing, at the same time, the state of the world which *in*-forms organism *A*

⇒ this change effects a modification of organism *A*'s behavior which is a description of its domain of interactions

⇒ the circle closes.

The modification of an organism's behavior specified by the signification of representations (in the autopoietic sense) is the *biological foundation of intentionality* as a property of a living system. Signification itself arises from the establishment of causal relations between an organism's different interactions (including interactions with representations), that is, from *experience*. In particular, it means that there may be no intentionality in the absence of signification, and since signification is a function whose argument is experience, and the amount and content of experience display a direct temporal dependency, the following conclusion is in place:

---

[16] For a discussion of some specific manifestations of reciprocal causality see Menary (2007).

Intentionality is the capacity of a living system to modify its state of reciprocal causality with the world on the basis of experience acquired with time for the purpose of sustaining the ecological system which enables reciprocal causality between the organism and the world

In other words, intentionality is a cognitive function of an organism as it is understood in autopoiesis:

> the functions that are usually considered cognitive are the result of a specialised subsystem of the organism continuously reconstructing patterns that are functional or referentially correlated with certain changes occurring in the environment. This set of patterns (built up during the existence of each cognitive organism) makes up what we usually call information (Moreno, Merelo and Etxeberria 1992: 67).

Such definition of information may seem to turn it into a 'technical' notion employed in computational modeling based on communication theory and the theory of information which deals with the more fundamental aspects of communication systems, such as the encoder and decoder and the problem of reproducing at one point either exactly or approximately a message selected at another point (Shannon 1948). However, the technical notion of information belongs to the observer's cognitive domain of descriptions where information is described (and the name 'information' assigned a semantic value) as an *externally manipulable thing* (hence the idea that a computer programmed by man can process information about cognitive states of a living system), while in fact it is determined by the structure of the nervous system and as such can be 'processed' only by this nervous system. A different organism (an observer) whose nervous system structure is similar (although never identical) to the structure of the observed organism because of a similar history of ontogenetic structural coupling (shared developmental experience), may 'process' *similar information* observing another organism's interactions (and this is what the biological function of mirror neurons appears to be), but it will never have direct access to the information 'processed' by the observed organism. The technical notion of information as a manipulable thing is applicable only to the environment as 'a set of processes and components that have to be recognized', and only partially (if at all) to the reconstructing patterns which correlate with changes in the environment. These continuously reconstructing patterns are those specific states of neuronal activity (representations) with which an organism interacts just because they correlate (are in a causal relation) with the changes in the environment; that is, they possess signification of which computers, for example, are incapable (and, consequently, incapable of cognition). This allows us to make a conclusion relevant not only to semiotics, but to the general theory of knowledge:

Representations as mental structures borne of experience are sign entities whose biological function is that an organism, by interacting with them, adapts to the medium by managing information

Some might object that intentionality as understood and described above stands in for what others call 'neural computation', and if so, what is to be gained by the introduction of the term which has been sufficiently compromised in philosophical theories of knowledge? To refute this kind of argument, I would suggest that since the notion of 'neural computation' is married to the technical notion of information as a manipulable thing external to the nervous system which does the 'processing', it is simply empirically untenable and, therefore, misleading. Intentionality viewed as a biological function grounded in experience allows to give an account of natural language as adaptive behavior and of linguistic signs as re-representing neural activity based in previous encounters that draw on sign-mediated events. This opens up *the developmental history of language* — if we drop the code (and, probably, computation - in the 'mathematical' sense) models.

3.5. Summary

The conclusion that the biological function of representations is adaptation to the medium by managing information, agrees very well with Linell's (2007) thesis that rather than being primarily engaged in modeling, imaging and imagining outside reality as the (first generation) cognitivist would have it, the brain is designed to control and monitor the interaction with the environment, and representations are there to serve the superordinate purposes of interaction with and intervention into the world. It also undermines the grand palace of orthodox linguistics built around the concept of sign in general and linguistic sign in particular. For truly, if natural (spoken) language is treated as a kind of adaptive behavior of an organism (a description of an organism's interactions with representations which are in a causal relation with the changes in the environment in its physical, biological, social and cultural dimensions), then linguistic signs appear to be not signs of environmental components but of representations which themselves are signs by definition. Consequently, language can be defined as *a system constituted by signs of signs* (cf. Peirce's "a sign for a sign" principle).

In turn, linguistic signs as a totality of environmental components constitutive of an organism's particular (communicative) behavior as a description of the niche of cognitive interactions, are subject to change to the extent that the organism and the medium are in a state of reciprocal causality. This explains, in particular, the historical change of language as its salient feature. In addition, since language is a specific medium (a subdomain of the cognitive domain of interactions), or part of the world, and at the same time an activity of which this medium is the product, its state is determined, in quite a natural way, by the state (the sum total of specific features) of the world, which helps understand the dynamics of the developmental history of human languages and their diversity.

Language is adaptive behavior which re-represents neural activity that is (partly) derived from linguistic experience. Cognitive dynamics use reciprocal causation and the mode of self-organization (intentionality) that shapes biological function. This fits Harris' (1981) principle of co-temporality and the distributionists' refusal to separate the linguistic from the non-linguistic. While interaction and languaging are biological, they are also a phenomenal dimension. Although words are, at root, sensory stimuli, they arise from bonding and mediating the components of structural coupling: they stabilize associations between things (the physical context) and phenomena in the world (the experienced world). Physical words, therefore, are phenomenological, indexical and serve an organism in self-constructing neural activity (representations) in interdependence with contexts. So (what we call) language is connotational. When we communicate, we draw on a history of the organism and the world's braiding together to establish resonance rooted in our individual histories of structural coupling.

Finally, in view of the fact that language is a dynamic semiotic dimension of cognition as it is understood in autopoiesis whose essential properties defy the coded equivalence principle cherished by traditional linguistics, it cannot and must not be regarded as a kind of code, whether digital or some other kind. Because "we live by metaphors", and because the code metaphor has been deeply ingrained in the minds of linguists since the time of the first cognitive revolution about 50 years ago (just like Turing's conceptual metaphor 'thinking is computation'), it is hard to let go of something so dearly familiar and closely nurtured. But let go we must, for the child has grown.

CHAPTER 4

SPEECH, WRITING, AND COGNITION: WHENCE THE COMMUNICATIVE DYSFUNCTION?

4.1. The Great Mystification

In Chapter 1, among the myths linguistics lives by, I mentioned the naïve linguist's belief, stemming from the objectification of language, that written language represented spoken language: graphic patterns on clay tablets, papyrus, parchment and, later, on paper were seen as objects 'containing' thoughts to be 'extracted' in the process of reading. This belief became the hallmark of official linguistic science almost at the very time of its conception as a scientific study of language and, strange as it may seem, has persisted to this day. However, while in the remote historical past such belief on the part of a linguistically unsophisticated language user might, probably, be perceived as quite understandable, the fact that today the written language bias should permeate linguistic science throughout really makes one wonder. It is as if evolutionary epistemology bypassed linguistics as a science leaving its conceptual apparatus unaffected by natural selection.

Speaking of selection, its primary role is to generate and maintain the reliability of our senses and cognitive mechanisms (including language), as well as the "fit" between those mechanisms and the world. Trial and error learning and the evolution of scientific theories are selection processes, and while in natural sciences these processes have played a decisive role, often accounting for radical revisions of the current scientific world view (suffice it to mention the pre-Copernican model of the universe as compared with contemporary cosmology, or the Democritian physics and contemporary nuclear physics, with quantum physics being a separate highly intriguing story), in linguistics it has never been the case. I foresee that the traditionalist might bring up, as a possible objection to this claim, the example of generative grammar often referred to as the Chomskian, or the cognitive, revolution in linguistics (Gardner 1985). However, in the context of our discussion of the relationship between speech, writing and cognition in this chapter, such objection must be refuted on the grounds that generative grammar itself is a stark example of the very written language bias in linguistics argued against in Chapter 1. Moreover, the unprecedented spread of generativism in the 1950s-1970s and its mesmerizing effect on cognitive science have been to a large degree responsible for categorical errors characteristic of first generation cognitive science briefly discussed in Chapter 2, such as that 'meaning is information', that 'thinking is computation', or that intelligent performance is symbolic processes in the head involving 'representations', to name but a few.

The whole issue of symbols and representations in the so-called 'mentalist' theory of cognition seems to derive from an unconscious metaphoric transfer, when our objectification of language, especially with the advent of writing, played a bad joke on us which we, regrettably, failed to see. Indeed, the relative ease with which a literate person can manipulate, when writing, graphic signs (words and phrases), combining and recombining them into strings of varying length and complexity (sentences and texts) to 'express' his thoughts which, in the process, become 'represented' in these signs, may well have been responsible for the clear analogy between the objectivist view of language as a manipulable thing 'out there' in the 'real' world, and the cognitive processes as mental 'representations' of that world 'in here', in the head, which are likewise manipulable things even though their nature remains somewhat obscure, to say the least. As it is, the very concept of information as a manipulable thing widely exploited in cognitive science seems to be grounded in the manipulability of linguistic signs which 'possess meaning': by combining and recombining linguistic signs (and corresponding meanings 'expressed' in them) we combine and recombine information. As argued in Chapter 3 (Section 4.3), the biological theory of cognition as the cornerstone of autopoiesis leaves no room for viewing information as an external manipulable thing for the obvious reason that two distinct organisms (cognizing agents, 'observers'), being structure determined systems, can 'process' only that 'information' which is determined by the structure of their respective nervous system. Because of a unique history of ontogenetic structural coupling in the case of each individual organism, what two (or more) cognizers 'process' and what we usually call 'information' can and does appear at times as very similar, especially when cognizers operate in a consensual domain of interactions, but, nevertheless, it is never one and the same 'objective' thing. It is much more appropriate to view information not as a thing, which is the standard approach in the theory of communication, but as a *process of indeterminacy reduction* (after all, the word *information* is a 'process-word', not a 'thing-word'). Specifically, in the biological theory of cognition information is understood as orientational activity modifying the behavior of the organism being oriented (Maturana 1983).

Remarkably, although not surprisingly, the theory of communication itself has been shaped under the overpowering influence of the original objectivist (that is, misguided) view of information as an external manipulable thing which possesses meaning. The paradigm example is none other than the written language itself. But if words don't *have* meanings in the traditional linguistic sense, if meaning, as proposed by Jordan Zlatev, is 'the relationship between an organism and its environment, determined by the value which particular environmental aspects hold for that organism', and if what we call (physical) words of natural language are just such environmental aspects among many other particular aspects, then for each particular organism the meaning of words would largely depend on the organism's environment (its niche, its *Umwelt*), or, to be more precise, on the organism's developmental history of fine structural coupling with the environment. But that is not yet the complete picture.

As has been argued in the previous chapters, just because an organism and the environment form a unity, not a union, the organism stands in a relation of *reciprocal causality* to the world. Human cognizers (particularly, their linguistic behavior) are components of the world with which each of us interacts, and as such they are also involved in this relation. In the biological theory of cognition, languaging has a biological function of orienting others in a consensual domain of interactions whereby they become in-formed in the environment. Orientation as a mode of adaptation occurs every time in a specific physical environment; it is, therefore, only natural to expect that aspects of the physical environment should affect linguistic behavior — from phonology to grammar. Because 'everything said is said by an observer to another observer', what is observed by different observers in different environments finds different manifestations in language(s). As Christopher Gauker (2003) suggests, the domain of discourse depends on what is relevant to the conversation in view of the situation in which the conversation takes place, so there is something about the environment that determines the content of the domain. According to Gauker, understanding on the part of the hearer involves both the meaning inferred from the speaker's sentences (which, it should be remembered, is not identical with the intended meaning) *and* the practical conditions of the environment. Not long ago, in his comments on the human art of translation on National Public Radio on December 14, 2006,[18] Andrei Codrescu made similar claims that the physical environment in which humans live was at least partially responsible for their language. As an example, consider the importance of an intricate system of indexicals for the vertical dimension in the languages of the peoples living in the highlands (Lawrence 1972; Gunajev 1977), and for the horizontal dimensions in the languages of the plainspeople (D. and K. Henry 1969; Sherer 1979), or the different degrees of vocalism in the dialects of Russian spoken on the plains in the European part and, for instance, in Siberia.

To make things even worse, linguists have had their share in creating the frustrating confusion of the written and spoken language by institutionalizing specific systems of notation in what they call 'linguistic analysis'. In a very insightful paper entitled "Analysis and notation: the case for a non-realist linguistics", Christopher Hutton (1993) discusses the conceptualizations of the relation between notational representations and their 'real world' analogues in the context of the quest for a realist philosophy of linguistics. Taking as an example generative linguistics as 'avowedly hypothetico-deductive and hence driven by a heuristic approach to

---

[18] Here's the link: http://www.npr.org/templates/story/story.php?storyId=6626808.

notation and analysis', Hutton convincingly shows how Chomsky's realism led, in effect, to a 'heuristic hocus-pocus', when the distinction between notation and analysis was maintained only by drawing on an as yet unrealized validation from neurophysiology. Basically, Hutton observes,

> linguists hold a variety of surrogational theories about analyses while being unable to prove the superiority of one school or methodology over another, which is a self-defeating state of affairs to say the least: verificationalism without verification. It is not that objections on the grounds of lack of verification are wrong, they are besides the point. Recognizing this involves understanding not only the limitations of notation systems, but also of their power in *conceptualizing language* [emphasis added. - A.K.] (Hutton 1993: 177).

The impact of notation-systems used in descriptive analysis on how language is ultimately conceived is something to be seriously taken into account if we want to avoid the pitfall of objectification, when an accepted system of notation and its 'objective' properties begin to be viewed as its analogue in the real world. As argued by Trevor Pateman (1987), there are real processes and cause and effect relations of a general nature specific to language which are often obscured by extraneous factors. Some of these processes and relations will be discussed in the subsequent sections in this chapter.

4.2. Speech, Thought, and Writing: Where Rationalists Go Wrong

4.2.1. Introductory Remarks

When we consider the current situation in linguistic science as far as comprehension of linguistic phenomena in their relationship to the cognitive domain of interactions of humans goes, we are back to square one in the controversy regarding the relationship between natural (spoken) and written language. On the one hand, the entire conceptual framework of traditional linguistics heavily depends on the objectification of natural language caused, provoked, or triggered by the seemingly objective, material properties of the signs of written language which are then extrapolated onto the signs of spoken language. The result is the 'holy cow' of the code model of language with all the numerous consequences that follow from the binary approach to linguistic communication. On the other hand, advances in contemporary cognitive science outlined in Chapters 2 and 3 disallow of the old conception of linguistic sign as a unity of material form (expression) and ideal content (meaning). Yet, the mainstream cognitive science continues to be in the grip of objective rationalism as regards conceptualization of language. Although, ontologically, spoken and written languages are essentially different, even though overlapping, subdomains of the cognitive domain of interactions of a specific linguistic community speaking a common language, phenomenologically they appear, at least to those who bother to ask the question, to be two realizations of one and the same thing. A typical example is the stance taken by Stevan Harnad in his article "Back to the oral tradition through skywriting at the speed of thought"[19] as part of the Internet seminar series *Rethinking Interdisciplinarity* held in 2003.

The goal of the seminar was to focus on the impact of new forms of communication on interdisciplinary research, and Harnad's paper, among other theme articles by distinguished scholars, caused a lively discussion. Because this paper explicitly addresses the relationship among spoken language, thought, and writing, with an angle on a new communication technology such as email, and because Stevan Harnad is an outstanding cognitive scientist whose views on language and cognition have been very influential, the paper can very well serve our purpose of trying to look beyond the trivial conceptualization of the above-mentioned relationship in an attempt to see, hopefully, what has been obscured by the strong rationalist tradition in the philosophy of language. It would be appropriate, therefore, to analyze the key propositions of Harnad's paper whose general content may be divided into two main sections dealing with spoken language and writing, joined together with a few claims about thought in its relationship to language.

4.2.2. Language

Harnad's propositions about language are as follows:

(i) Language gave humans the adaptive advantage (in the terms of survival and reproduction) by making possible the acquisition of knowledge by 'hearsay' ('symbolic theft') rather than the hard way, i. e. from experience. Most of our knowledge is recombinatory: the child grounds his first word meanings through direct sensorimotor toil and can then (in principle) acquire all the rest through symbolic theft, consisting of recombinations of his already-grounded symbols, rather as in the case of dictionary definitions.

(ii) Language is a form of reciprocal altruism, and the oral tradition arose out of this reciprocal altruism: except if we are competing for the same finite resource, you lose nothing if you tell me, honestly, what is and is not edible, or which are and are not dangerous predators; and tomorrow it may be I who knows something that you do not. So "cognitive barter" may be a better descriptor than "theft" for the adaptive advantage conferred by language. It can be thought of as a collective, serial form of cognitive barter, whereby we inherit the knowledge of those who have it already, and in re-

[19] Here's the link to the paper: http://www.interdisciplines.org/defispublicationweb/papers/6.

turn, we add what we ourselves know, or at least pass on what we have learned. It is the oral tradition that allowed the benefits of language not only to be collected and passed back and forth within a generation, but to be passed on across generations.

(iii) Language actually began in the praxic modality of body movement, gesture and imitation, yet it is neither an instrumental connection nor a morphological resemblance to its referent that makes a symbol a symbol. It is the fact that we intend it to refer to that referent. This is Saussure's "arbitraire du signe": although the power of naming is more likely to be first discovered in an instrumental or imitative context, the instrumentality and the mimesis eventually become irrelevant to the naming, and the name itself might just as well have been in binary code, once it is used in discourse. Language is in this sense inherently digital.

(iv) The true advantages of language do not come from naming, but from combining and recombining strings of names into propositions that define or describe further truths. The optimal sending and receiving medium would need to be a faster and more independent one than bodily gesture — a medium one can use even if one's hands are full, or if one's interlocutor is not watching, or in the dark. In short, once the advantages of language were discovered and being used, the oral medium was the natural one to specialize for this purpose, and so it did.

(v) People are born with a pre-adaptation to be able to learn language very quickly, and a strong predisposition to do so. Hence we have evolved brains that are born "language-prepared".

At first sight, all the above claims seem to be well justified. Indeed, language gave us the adaptive advantage we enjoy so much, even though people from the street do not think about it in such terms. But the layman would agree that language is used to 'acquire knowledge', and this belief is the epistemological foundation for the concept of public education in modern societies: educational institutions, such as schools, colleges, and universities, are special places where (young) people learn 'the easy way', by 'hearsay' or 'symbolic theft', as Harnad puts it. The belief that schooling prepares young people for mature social life by letting them 'acquire' the knowledge conveniently accumulated and stored in language the 'easy way', without any first-person experience, still holds strong among laypeople, although professional educators might give them a very different piece of mind. It has become increasingly obvious over the past decades that high school graduates, for example, show alarming ineptitude in using the so-called acquired knowledge to their benefit when addressing practical tasks in their everyday public life. This inability to turn 'acquired knowledge' into an advantage casts a shadow of doubt on Harnad's claim that languaging as a process of knowledge acquisition, or at least the greater part thereof, occurs outside first-person experience. True, he makes a concession by admitting that the child grounds his first word meanings through direct sensorimotor toil. Once it has been done, says Harnad, the child can ('in principle') acquire all the rest through 'symbolic theft', consisting of recombinations of his already-grounded symbols, rather as in the case of dictionary definitions of words which are defined in terms of other words: "once those words (elementary symbols) — have been "grounded" directly through honest toil, the good old way, all the rest of the dictionary (and encyclopedia, and every other kind of text, written or spoken) can then in principle be picked up by hearsay, recombinatory hearsay, composed out of the symbols we have already grounded directly through experience" (Harnad 2003). This is what the cognitive scientist Stevan Harnad honestly believes, and in this he is not alone — as already mentioned, this is the kind of epistemological premise on which the traditional philosophy of education has built its ideology. And this is where both go wrong — at least, as far as 'symbolic theft' goes.

Harnad's notion of symbolic theft is based on an assumption that symbols (that is, linguistic signs) possess objective meanings, meanings which have the same 'value' for any language user regardless of any other 'extraneous' factors such as all the multifarious aspects of the physical environment in which the 'theft' occurs, or the developmental histories of the sign user and the 'thief' which account for their unique characteristics as structure determined living systems. As has been argued in the previous chapters, meanings are not contained in signs, and although we are all much used to the 'container metaphor' exploited in linguistic descriptions, the fact itself does not make this metaphor any more real than our trivial use of the expression *The sun rises in the east* makes real the geocentric model of the universe. Speaking of 'symbolic theft', Harnad commits the rationalist's error of reification, because he views linguistic signs as objective things existing independent of the people who (might) use them. More than that, he speaks of elementary symbols (words learned the 'hard way'), thus implying the existence of complex symbols as combinations and recombinations of the elementary symbols. What better illustration for the Two Dogmas of Reificatory Semantics could one offer?

However, as a kind of biological (adaptive) behavior predetermined by an organism's history of fine structural coupling with the environment, languaging cannot be interpreted other than within the context of the organism-environment system. Such interpretation can never be one hundred per cent accurate for reasons discussed in Chapter 3. That is why we simply cannot speak of our understanding language as 'symbolic theft' just as we cannot speak of a magpie picking a bright bead or a dropped ear-ring in terms of 'jewelry theft'. To prove that this is not a mere speculation, I invite the reader to recall his school-time experience when a particular text to be studied for a particular class resisted the reader's comprehension regardless of the time and effort put into the preparation for that class (and, of course, for different readers they would be different classes with varying degrees of comprehension problems). Why should this happen at all if all knowledge 'expressed' in language is, according to Harnad, recombinatory, and the text is written in the reader's native tongue? Here Lev Vygotsky's *Thinking and Speech* comes to mind, where he warned against bare verbalism in teaching when thoughtless acquisition of new words, supplanting *living knowledge*, didn't go any farther than simulate corresponding concepts which the child, in fact, lacked, concealing thereby the actual vacuousness of the mind.

As yet another example, consider people (we all have met them in our life) who never read certain types of books usually labeled by literary critics as 'serious literature' (not to mention dictionaries or encyclopedias) while being voracious readers of detective stories or paperback fiction. Why, having read all the books by Sydney Sheldon or John Grisham, such a person may have never read (at least, to the end) a book by James Joyce or William Faulkner? The answer is, because such reading exerts a lot of strain on this particular kind of reader as the asserted recombination of knowledge contained in linguistic signs *does not work for them*, no matter how hard they may try. Here Harnad's ingenuity must be praised and admired, for when speaking about the child acquiring the rest of knowledge through 'symbolic theft', he leaves a clever way for retreat by using an inconspicuous bracketed expression, *in principle.* Now, I don't think Stevan Harnad is the kind of person to make light of the words he uses. By making this reservation, he seems (even be it unwittingly) to imply that, as far as available data go, 'symbolic theft' as a cognitive phenomenon underlying linguistic communication has not been empirically validated and remains a purely hypothetical assumption. As a matter of fact, the notion of 'symbolic theft' (or knowledge transfer, in a more common parlance) is precisely what is the problem with the concept of literacy characteristic of the mid-twentieth century which, as many in the field recognize today, is in need of dramatic revision (more on this in the next section).

Harnad's second proposition about language is that it is a form of reciprocal altruism out of which arose the oral tradition. With this in mind, he suggests the term "cognitive barter" as a better descriptor than "theft" for the adaptive advantage conferred by language (remember, this advantage is the ability to acquire knowledge by 'hearsay'). The notion of reciprocity, as argued in Chapter 3, is very important in understanding languaging as adaptive behavior, and Harnad cannot and must not be refuted here on this point. At the same time, he connects reciprocity with altruism: a person loses nothing if he tells another person what is and is not edible — *except if they are competing for the same finite resource*. I may not be an expert on the issue of language evolution, but surely at the time when language evolved (if evolve it did), which was many millennia ago, food resources were hardly infinite — not on the global scale, but on the scale of a particular habitat of a particular tribe. Nomadic life many peoples live even today, not to speak of the times long gone, is, in fact, the direct consequence of finite resources (the effect and cause relationship). Moreover, to assume that oral language arose from altruism would be to underplay the importance of the primary instincts for survival and reproduction which, in the case of a single organism, override everything else. Of course, we, humans, like many other species, are social animals for whom survival and reproduction have become a biological function of their social organization. From this point of view, we should agree with Harnad that languaging can be thought of as a collective form of cognitive barter, but to agree that, thereby, we 'inherit' the knowledge of those who have it already and, in return, add what we ourselves know, would be a rash thing to do. It is alright for a first generation cognitive scientist to believe that knowledge contained in language is passed 'back and forth' within a generation and, what's more, passed on across generations. But for a second generation cognitive scientist, who sees knowledge as something being actively constructed every time we engage in communication viewed as interactional activity with orientational effect, rather than as something being 'extracted' from linguistic signs, such approach is clearly untenable.

Suppose we accept the 'barter metaphor' in describing linguistic communication while rejecting the idea that it is knowledge which is the object of barter. What, then, would be an adequate alternative for explaining linguistic communication? We can avoid the reification fallacy if we concede that, as 'cognitive barter', linguistic behavior consists in our exerting orientational effects on the linguistic behavior of others which is in a relationship of *reciprocal causality* to our own linguistic behavior (see Chapter 3, Section 4 for a detailed discussion). The next proposition Harnad makes, namely, that linguistic signs are arbitrary, is not new. Importantly, he admits that the power of naming was more likely to be first discovered in an instrumental or imitative context. This agrees well with the findings made by psychologists, such as Alexander Luria, who stressed that at the earlier stages of ontogenetic development "the word is woven into a situation, gesture, facial expression, intonation, and only on this condition it acquires its referential properties" (Luria 1998: 69). It also does agree with current research in semiotics (see, for example, Pajunen 2002; Zlatev, Persson and Gärdenfors 2005). Yet he dismisses this circumstance as irrelevant to the naming, suggesting that the name itself might just as well have been in binary code, once it is used in discourse. In this sense, language is inherently digital. Now, what should one make of this?

Surely, if we start from the assumption that language, in the abstract, is indeed a code (in the meaning 'cipher'), then it wouldn't really matter what particular encryption technique is used for converting information or material in one form into a representation in another form. But for our initial assumption to have any empirical value, we must, as Nigel Love (2004) argues, ask the question, "What is it that language, as a code, encodes?" As was illustrated in Chapter 1 (Section 5) by the argument offered by Love, we cannot, really, identify a realm of non-linguistic thoughts or ideas that language, according to the rationalist view, encodes. To reinforce his argument, Love goes on to address the related problem of identifiability of linguistic units, and, quite convincingly, takes apart the idea that when we produce an utterance there is some definite linguistic entity over and above our utterance itself, of which the utterance is a repeatable instance (a unit of Saussure's *langue*, the so-called 'invariant'). This leads him to the following conclusion: "It is hard to make sense of the idea of a code if it turns out to be impossible to identify unambiguously the operative units of which the code allegedly consists" (Love 2004: 540). I find this conclusion hard to disagree with. Moreover, as I have argued elsewhere (Kravchenko 2003a), the approach itself to linguistic units as stable objects is based on what Chris Sinha (1999) calls the Dogma of Autonomy, and because of that it cannot be illuminating or elucidating. That the approach to language as an autonomous system is unproductive was pointed out by Bertrand Russell (1948), who wondered why some philosophers attributed to language the feature of autonomy forgetting that language was meant to relate to facts and facilitate communication with the world. The untenability of such approach was very well summed up by Boris Gasparov:

> Everything that one could, and what linguistics is apt to readily and willingly accept as stable linguistic objects, that is, all the "end products" of linguistic activity which the speaker produces himself and which he receives from other speakers, are really such only upon a very cursory glance. Any such product does not exist for the speaker other than in a continuum of associations, intellectual and emotional responses, spontaneous recollections, intuitive and conscious assessments of the situation and the protagonist, and, following such assessments, anticipatory predictions as to how the situation of interaction with the protagonist is going to develop afterwards and what new communicative tasks stem from this. The properties of such a product (conversation remark, spoken utterance, written text), what it appears to be from the point of view of a given speaker in a given situation, are inalienable from the continuously flowing medium of multidirectional linguistic activities, into which our "product" is immersed and the flow of which, in its turn, it changes just by the mere fact of its immersion therein. Any instance of language use, be it a highly valued piece or a passing remark in a conversation, is of itself a corpuscle of the ever moving flux of human experience (Gasparov 1996: 9 ff).

During the on-line discussion of Harnad's paper it was pointed out that the idea of binarism is applicable only to units of an artificial sign system, and in that respect writing *is* a code of a kind, and this is exactly what makes it so different from spoken language which is not digital because it is not denotational, but connotational (I refer the reader to the relevant sections in the previous chapters). However, Harnad insisted that speech and writing 'are in 1-to-1 correspondence (at the lexical, 'word/object' level)' because 'language is most definitely denotational if anything is' ("snow" certainly denotes snow). As for binarism, his argument was that he could have named all his children in Morse code if he had wished. Although he, in fact, hadn't (I wonder why?), he could still transcribe their names in Morse code, if he liked, and even say them that way (and, again, I wonder if Harnad ever tried to do that in earnest). Regrettably, my counterargument that the fact that I can walk on stilts to cross a pool of water does not make my legs instances of stilts, or that my ability to walk on hands doesn't make them instances of legs as a locomotive apparatus, either, didn't seem to have any weight.

Overall, Harnad's position on the issue of arbitrariness of sign is not surprising, especially in view of the fact that, to him, *semiosis* (the sign generation process), just like *autopoiesis*, is a 'cult word' which 'means many things to many people, with the intersection empty or trivial', although one cannot help noticing that, for Harnad, the word *semiotics* doesn't seem to belong to that category for reasons which remain unknown. As a matter of fact, there is hardly a word in, for example, the English language, which doesn't mean many things to many people, and there is a notorious lack of consensus among linguists themselves on the meaning of most of the words they use to refer to the core concepts of what they call 'linguistic science' (ready examples would be such terms as *word*, *sentence*, *meaning* and, of course, *language*). The belief that language is denotational makes Harnad, just like many rationalists before him, claim that words possess meanings, that these meaning are objects, and this is exactly where the problem is with orthodox thinking about language discussed in Chapter 3.

Harnad dismisses findings, provided by research in semiotics, about the importance of the instrumental or imitative context in which, as he acknowledges, the power of naming was more likely to be first discovered. But what would a cognitivist, such as Stevan Harnad, make of very strong iconicity, as in large-scale ideophonic systems in languages lacking fusional morphology? Moreover, as pointed out by Jess Tauber (personal correspondence), the iconic principle may not necessarily involve phonology. Thus, in certain languages complex word stems containing instrument prefixes, roots, and locational/path suffixes seem to end up having some sort of dynamic parallel organization to the human body itself. In the same slot as instrument prefixes are usually bodypart terms, or at least typical actions of standard parts. Interestingly, these are almost all upper body — we just don't do that much instrumentally with our lower half worth lexicalizing into the system, so it would seem. As for the pathway/locational suffixes, these of necessity deal nearly entirely with the lower body and its contacts with the ground, etc.

According to Tauber, in almost all the languages that have these two subsystems (languages of western North America, in families such as Siouan, Uto-Aztecan, Hokan, Penutian, etc.), they are on opposite sides of the main word root.

Examination of the root system itself shows that it often deals with processes and states that might typically be found *inside* the human body, or on the outside on the trunk. This is rather different from other language types, where instrument and path notions might more typically be part of the inherent meaning of the root itself. There are also languages where the order is reversed — for instance, ancient Indo-European and Kartvelian. There are, in addition, languages which seem to invert the notions — so, for instance, upper body parts are now patientive, not instrument like, and pathway terms are now interpreted as pseudo-agentive. These also come in two orderings. So, given this typology, Tauber argues, one can cover a *lot* of languages, so there must be some sort of underlying 'pressure' to create this sort of stem structuring. A similar point is emphasized by Didier Bottineau (in press):

> [A]lthough human language is widely documented to be unmotivated, there tends to emerge locally remarkably organized subsystems, islands of order in oceans of chaos, which are strongly suggestive of an underlying ordering principle.

Apart from anything else, data of the kind mentioned by Tauber seem to provide strong evidence that language is not an arbitrary symbolic code, after all.

Now, let us go on to the 'true advantages of language' as seen by Stevan Harnad. That the oral medium was the natural one to specialize for signaling in the case of not only humans, but most higher order animals as well, is hardly a matter of dispute. In fact, this aspect of semiosis (am I being cultist?) was discussed in one of my earlier papers (Kravchenko 2000). But do the advantages of language really consist in 'combining and recombining strings of names into propositions that define or describe further truths'?

As one from the rationalist camp, Harnad obviously belives that it is possible to define meaning on purely logical (formal) grounds, which is the trademark of first generation cognitive science based on the computational metaphor. However, and this is a well known fact about the rationalist philosophy of language, such attempts have been infallibly countered by the fortitude of the problem: no matter how hard and with what perseverance adepts of truth semantics tried to resolve it, their efforts have invariably turned futile. The explanatory inadequacy of such approach has been discussed at length in the literature (Dummet 1981; Jackendoff 1983; Seuren 1985, 1998; Fillmore 1986; Lakoff 1987; Ludlow 1993; Gärdenfors 1999), so I am not going to dwell on that and will offer just a few comments on the issue I have previously made elsewhere (Kravchenko 2003a).

The point of the matter is that "being true" as a propositional feature of correspondence to a given state of affairs is quite straightforwardly determined by the cognitive context in which the proposition is made. Specifically, account should be taken of the extent to which the content of an utterance is influenced by an individual speaker's sensations and feelings; that is, account should be taken of the observer's viewpoint (Kravchenko 1992). The sophists already had this understanding: they pointed out that there are such elements of thought which are not overtly expressed in an utterance, nor can they be inferred from it, yet they participate in the formation of meaning and truth conditions.

The rational paradigm in linguistic semantics rests on a fundamental premise that knowledge, at least as it is reflected in language, is constituted by an explicable set of logical operations and procedures which make it possible to determine the truth or falsity of propositions as situation structures expressed in what is known as full linguistic signs, that is, sentences/utterances (Larson and Segal 1995; Platts 1997). Seen from such a perspective, semantics as the study of meaning finds itself on a par with formal logic and mathematics since they all imply logical calculus (Tarski 1956; Seuren 1998; Keenan and Stabler 2000 *inter alia*). If such a position is taken, then to *know something* signifies "to have sufficient reasons to believe that a given statement is true/false." But what is the criterion used to determine whether such reasons are, in fact, sufficient? The same old *ratio*, because outside it the principles of formal logic are inapplicable since they rest on a given set of initial axioms. This justifies Ray Jackendoff's call to re-evaluate the traditional preoccupation of semantics with explicating the notion of the truth or falsity of a sentence, as semantics must be studied not from the point of view of logical relationships but from the point of view of the structure of the human organism; that is, a proper theory of word meanings must be richer than standard logical models. As Jackendoff (1997: 543) points out, the traditional Tarskian definition of truth:

> *Sentence* S *of language* L *is true iff conditions* $C_1, \ldots, C_n$ *obtain in the world*

> must be reinterpreted as an issue of judgment:

> *A speaker of language* L *judges sentence* S *true iff conditions* $C_1, \ldots, C_n$ *obtain in his or her construal of the world (subject to limitations of memory and processing, and under an idealization of uniformity among speakers).*

The plight of traditional linguistics is that it has been literally trapped in a vicious circle in its quest for a resolution of the paradox it itself created by separating language from the world and opposing one to the other. The intriguing part, however, is that the reasons for such opposition have never been consistently and convincingly explained.

Harnad's final proposition about language, namely, that people are born with a pre-adaptation to be able to learn language very quickly, and a strong predisposition to do so, or, in other words, that the human brain is "language-prepared", doesn't cause any objections. Except, probably, one fine point concerning the meaning of the expression "to learn language". As, for Harnad, language is primarily a system of arbitrary symbols used to code truths by combining and recombining names into propositions, the meaning of the expression "to learn language" must mean something like 'memorize all the symbols available in dictionaries and encyclopedias and their meanings (objects they stand for), as well as the allowed possible combinations between any two (combinations of) symbols, in order to be able to combine and recombine them to express truths'. Which brings us back to the original Cartesian error (see Chapter 2) and the positing of a special realm of truths as mental content to be expressed in signs which are 'out there' (the ubiquitous 'container metaphor'). But even Charles Ogden and Ivor Richards themselves emphasized in *The Meaning of Meaning* the experiential aspect of symbols we acquire when learning language:

> ...It is actually through their occurrence together with things, their linkage with them in a 'context' that symbols come to play that important part in our life which has rendered them not only a legitimate object of wonder but the source of all our power over the external world (Ogden and Richards 1923/1927: 34).

By attributing this feature to symbols only at the early stage of ontogenetic development, Harnad created what has become known as the 'symbol grounding problem' (Harnad 1990) which, as a special workshop in Plymouth in 2006[20] showed, doesn't seem to have a solution — at least, as long as 'arbitrariness of sign' continues to remain an unquestionable axiom.

On balance, Harnad's view of language is a very good illustration of the grip that the axioms of orthodox linguistics hold on first generation cognitive science with its Cartesian methodology, which is unwilling to trade the holy cow of rationalism for any alternative method in the study of language, even if current cutting-edge research, both theoretical and empirical (acknowledged by Harnad *en passant*), has shown that such an alternative may be a far more promising and gratifying enterprise in the attempt to understand what language is and what language does for human cognizers.

4.2.3. Language and Thought

As I hope to have shown in the above discussion, the main point in Harnad's approach to language as a cognitive dimension of human activity is that there are two distinct realms, one of thought ('propositions describing or defining truths'), and one of language (combinations of arbitrary symbols used to express proposi-

[20] Here's the link to the workshop's webpage: http://www.tech.plym.ac.uk/SoCCE/ESG2006/.

tions). So it is interesting to see how well it stands with Harnad's propositions about thought which we find in his article. Actually, there are two of them:

(vi) The speed of human thought is of approximately the same order of magnitude as the speed of speech, because if we thought much faster than we speak, we would encounter short-term memory and interference effects when we tried to vocalize our thoughts. A more fundamental constraint on the speed of thought is that discourse is interactive. Originally, there were no monologues, only dialogues, in which the cognitive interaction was bilateral, synchronous in real time when the information exchanged was of some immediate and, eventually, practical benefit to one or both parties. That was the pragmatic context in which language acquired and exhibited its adaptive value.

(vii)Dialogue not only informs, and trades existing information; it also elicits and stimulates novel interdigitating thoughts that may not have occurred solo. In that sense, interactive cognition may create information. Language is already recombinatory: combining the resources of two interacting heads opens up possibilities that go beyond the sum of the two parts thinking alone. Thinking itself co-evolved with speaking, not only in tempo, but in its sequential and recombinatory nature and its interactivity. So much of thought is inner dialogue, and even when an inner monologue, addressed to some imaginary or remembered interlocutor.

Here, again, Harnad displays a mix of beliefs rooted in two different approaches to language: as a cognitive activity, and as a system of symbols for expressing thought. On the one hand, because discourse is interactive, language is dialogical: 'originally, there were no monologues'. This is fine, and I don't think anyone will contest that. But on the other hand, 'the speed of human thought is of approximately the same order of magnitude as the speed of speech', because 'if we thought much faster than we speak, we would encounter short-term memory and interference effects when we tried to vocalize our thoughts'. What to make of this statement?

The obvious inference is that we think in words. However, in the ensuing discussion Harnad conceded that we do not think only in words, but words are 'the main way (though not the only way) that we express our thoughts to one another'. Now, the idea that we use language to express our thoughts to others is the classical view on language much criticized in the literature as the 'language myth' (see Chapter 1). Harnad speaks of some immediate benefit to the parties of the information exchanged in the course of the cognitive interaction which is bilateral and synchronous in real time: 'dialogue trades existing information'. As is clear from the previous discussion, for Harnad 'exchange of information' is expression of thoughts (propositions) by combining and recombining symbols which possess meaning. However, this creates an insoluble controversy: if, originally, there were no monologues, and if thought is something expressed in (dialogical) speech, then it would appear that there were, in fact, no thoughts to express and no existing information to trade until interlocutors engaged in discourse which is the expression of thoughts!

However, the cognitive part of Harnad the scientist as compared to the rationalist part (or should I call it the 'better half'?) cannot be subdued, and he admits that dialogue 'also elicits and stimulates novel interdigitating thoughts that may not have occurred solo. In that sense, interactive cognition may create information.' This is a relief, but what is meant by 'information' in this case? Harnad is explicit: it is 'combining the resources of two interacting heads'. But he stops short of explaining what he actually means by the word 'resources', obviously guided by the rationalist belief that words (linguistic symbols) are autonomous independent things which possess meanings; the latter are objective in the sense that they do not pose interpretation problems for individual symbol users. And we all know that this is not so. So what could Harnad mean by 'resources of interacting heads'? And do heads as structural parts of our bodies really interact, and if they do, how do they do it? I can think of one instance of such interaction when, in certain cultures, people rub noses as a form of greeting. But surely Harnad doesn't mean that. As was mentioned in Section 5 of Chapter 2, to him cognition is autonomous activity (internal process) that generates input/output capacity, and it takes place 'entirely within the brains of cognizers'. But what, then, about information which exists prior to dialogical interaction? If it is a thing — and it sure seems to be such, or there wouldn't be any talk of 'exchanging' it — then it is either an input or an output (oh, the pervasiveness of the 'container metaphor'!), either of which would have to be in linguistic form because information is 'propositions'. Personally, I don't see a way out of this solipsistic vicious circle.

Fortunately, Harnad makes a much more interesting suggestion, although in the form of timid acquiescence, that 'interactive cognition may create information'. This puts a non-Cartesian cognitive scientist such as myself on more familiar turf inasmuch as information created by interactive cognition is, to him, orientational effects on one another caused by the cognizers' (linguistic) behavior which has meaning — the relationship between an organism and its environment, determined by the value which particular environmental aspects hold for that organism. From this point of view, 'created information' that Harnad speaks about may be interpreted as *meaning* arising from the cognitive effort to evaluate the other party's (linguistic) behavior as an environmental aspect. And because meaning is a cognitive function of an organism's interaction with the environment, there can be no signs which exist prior to meaning (see Chapter 3, Section 3), be they arbitrary, non-arbitrary, or whatever. A sign is always a sign to *someone*, and if this condition doesn't hold, if there isn't a 'someone' to whom something stands in for something on the basis of that someone's experience of interactions with both the entity we call 'sign' and what it stands for (its referent) as relevant aspects of the environment, any talk about there being symbols 'out there' to be learned, becomes pointless. Yet this is precisely what rationalists insist on.

Another claim made by Harnad makes one feel uneasy, the claim that 'thinking co-evolved with speaking, not only in tempo, but in its sequential and recombinatory nature and its interactivity'. On the one hand, human ontogeny seems to provide solid proof for that, and Harnad's assertion sounds very Vygotskian, to say the least. But on the other hand, it is not quite clear in what relation thought stands to cognition. If *thought* is synonymous to *cognition* (although we know that synonymy is a relative notion and there are no true synonyms), and 'cognition takes place entirely within the brains of cognizers' as claimed by Harnad and Dror (see Chapter 2 for discussion), then speech as something external to the brain (specific acoustic phenomena that accompany human interactive behavior) must be viewed as either existentially preceding thought ('input'), or following it ('output'). Now, because 'thinking co-evolved with speaking', and oral speech evolved for the speed of 'interactive' thought, as Harnad argued during the discussion, there must be some sort of cause-effect relationship between them: in the first case, speech would relate to thought as cause to effect, and in the second case, the cause would be thought and speech the effect. And of course neither proposition is plausible enough as they bring us back to the vicious circle mentioned above — that is, unless we adopt a distributed view of language and cognition argued for in Chapter 2. Moreover, 'co-evolution' of thought and speech poses the question of thinking without speaking as part of our general cognitive ability which we share with many higher order animals. And here the 'speed of thought' thesis doesn't seem to hold at all because we often think much faster than we speak, and we do it in the 'monological mode', so to speak. Human intuition, sudden insights, revelations and the like all seem to be relatively independent of (the speed of) speech, be it external or internal. By insisting that our thinking must remain in phase with the speed of speech if we want to communicate in parallel, Harnad leaves open the question of what goes on in our heads when we are not engaged in communication. Surely he doesn't think that we stop thinking when we are alone?

But what if *thought* were not synonymous to *cognition* while being in a reciprocally causal relationship to speech? By acknowledging the dialogical (interactive) nature of thought, Harnad takes it beyond the skull of an individual head, thus separating it from cognition. At the same time, such approach would allow granting 'cognizer status' to organisms without speech — what, then, would be the difference between a human cognizer and, say, a chimp or a dog? The answer is obvious: language (speech). The position taken by Harnad is riddled with baffling contradictions which the methodology of rationalism seems unable to resolve.

4.2.4. Writing

Like his view on spoken language and thought, Harnad's views on writing are also multidimensional and may be summed up as follows.

(viii) The advent of writing brought both advantages and disadvantages. The advantages are that writing leaves a (potentially) permanent record, guaranteeing continuity. Writing made science and scholarship if not possible, then at least far more likely. The negative effect of writing on the temporal, interactive dimension of linguistic discourse was in transforming it into asynchronous, off-line monologues instead of the synchronous dialogues for which our brains and our thinking capacity are optimized. Writing is asynchronous discourse — out of phase with the speed of thought and of synchronous mental interaction. But because its benefits were so great (and because the oral modality was still there), literacy was a benefit and a revolutionary advance, especially when its reach was greatly enhanced by the invention of the printing press. However, both writing or the invention of print were purely cultural developments which didn't cause any compensatory organic change in our brain. And writing seemed, by its very nature, to entail dissociating asynchronous written interactions from the speed and synchrony of interacting thoughts.

Among the propositions that cause absolutely no objections are those about the role of writing in sustaining continuity and making science possible. As for science, and as mentioned in Chapter 1, it is itself a construct of language, specifically, a construct of written language, which fact makes it perfectly sensible to speak of the semantics of science (Kertész 2004). But this is not among the issues that are in the focus of our attention, so we'll leave it at that and turn to writing.

What is the evolutionary purport of writing, if any? It helps overcome the spatio-temporal limitations of spoken language by establishing an atemporal (as contrasted with the temporal and local) existential/informational environmental medium for humans. As Harnad puts it, *Verba volant, scripta manent.* The medium of natural language is akin to the human genome in biology, since it is only in and through this medium that we become humans (for the discussion of some related points see Chapter 2, Section 5). Writing, especially after the invention of the printing press, heralded a new level of organization of the cognitive domain of interactions as the bonding substance of the 'organism' of a community of individuals. Over time, the domain of written texts became, at least in developed cultures, a kind of Petri dish for culturing individual humans, although with one essential difference: by thriving in the 'nourishing' medium of written texts, humans, far from depleting it, grow and cultivate it by forever adding new texts intricately interrelated with the already existing ones through humans' interactions with them (hence the continuity mentioned by Harnad). However, there's a catch: by being unable to cognitively thrive in a written culture, or downright refusing to do so, an individual human does not thereby commit an existential suicide as a living organism. It is his potential for social adaptation which is put in double jeopardy: inability to effectively interact in the domain of written texts turns out to be a major cognitive dysfunction affecting the overall power for adaptation throughout the organism's developmental history (more on this in the subsequent sections). Literacy, being undoubtedly a benefit and a revolutionary advance at the time, cannot of itself be regarded as something that guarantees cognitive development in contemporary post-industrial society. It is a tool, and a very effective one, but a tool, no matter how advanced and effective, in the hands of an unskilled user may just as well turn into something dangerous, or even deadly — and the number of people committing cognitive/communicative suicide daily is simply astounding. Some of the reasons for that will be discussed shortly.

Meanwhile, I would like to focus on a peculiar inconsistency in Harnad's treatment of spoken and written language. He observes that 'writing or the invention of print were purely cultural developments which didn't cause any compensatory organic change in our brain', and I have no gripe about that. But the question is, "If spoken language is a natural code, and written language is a cultural code which stands to spoken language in one-to-one correspondence, where is the divide between the two?" Harnad sees it in that writing is asynchronous discourse which is 'out of phase with the speed of thought and of synchronous mental interaction'. That it is out of phase with synchronous mental interaction and, at least partially, with the speed of thought, is self-evident. Less evident is the rationalist error which Harnad makes when he insists that language as a symbolic code 'gave humans the adaptive advantage by making possible the acquisition of knowledge'. Not repeating the anti-rationalist arguments given in the previous subsection, I would like to note that if oral language serves the function of knowledge acquisition by being a digital code, as Harnad would have it, it seems that the word *code* is used by him as a description of a specific aspect of the human biological makeup — a function of the brain, for example, not unlike the function of DNA in the human genome. But written language is also a code which doesn't essentially differ from the oral language code inasmuch as knowledge acquisition is involved, yet it is a cultural artifact and as such cannot be viewed as having any organic basis in the brain. However, in Harnad's interpretation, this does not affect the ability of written language to be used to acquire knowledge. And if we agree that such ability is really an adaptive advantage, the inevitable inference is that there must be no difference between the kinds of knowledge acquired with the help of speech on the one hand, and with the help of written texts, on the other. And this is simply not true.

When Harnad says that writing, by its very nature, entails dissociating asynchronous written interactions from the speed and synchrony of interacting thoughts, he forgets (or just doesn't bother) to mention that thoughts, as part of what constitutes cognition, happen entirely inside our heads, and for that reason there is, really, no way they could 'interact'. This is a metaphor, of course, but what is behind this metaphor? It is, primarily and overwhelmingly, our linguistic behavior or languaging — a specific kind of cognitive activity integrated in the complex dynamics of multimodal human interactions with the environment. And while in the case of oral communication speaking humans interact with other humans in real time and space, in the case of so-called written communication there is, in fact, no such interaction.

Viewed as a kind of dynamic process that occurs in real time and space, writing is interaction with physical lifeless objects such as a writing implement and the material on which markings are left. In retrospect, writing as a process is a reflection of the writer's interaction with his own thoughts, and on this view we may speak of interactive thought, but still it is very different from the dialogical interaction in oral discourse. Viewed as a static artifact, a written text becomes the object of interaction for the reader, but that does not imply that the reader interacts with the thoughts of the writer presumably stored in written signs. Enthusiastically describing the advantages of email as 'skywriting' ('because it can be one-to-many in almost-real time, as if the email were being written in the sky for all to see'), Harnad goes as far as to claim that 'this instant quote/commenting capability can even restore to digital interactions with inert texts (even when their authors are long-since dead) some of the "live" interactivity of the oral tradition — albeit rather one-sided in the case of an expired author, but other skyreaders can in principle take up the interactive baton, and it can be rather exhilarating to carry on a live if unilateral dialogue with a long-dead author in almost real-time before a live audience that is potentially the entire planet'. The notion of 'unilateral dialogue' as 'one-sided interaction' is something hard to comprehend, and I am not even going to try. But it seems that what Harnad really means is not interaction with an expired author whose thoughts have been 'expressed' and 'preserved' in a particular text, but, rather, it is interaction of the reader with his own cognizing linguistic self. What the reader interacts with is his own interpretations of the signs he visually perceives, and these interpretations, in turn, depend crucially on the reader's life-time experience of linguistic interactions, both oral and written. There may be a profound lack of relevant experience caused by various circumstances, resulting in the potential reader's inability to 'extract' any knowledge whatsoever. There is a saying in Russian about such pseudo-literate folk: *Smotrit v knigu, a vidit figu* which translates roughly as '(He) Looks in the book and sees nothing but a fig'.

For Harnad, email embodies the best of both the oral and the written traditions: it is potentially almost as fast as synchronous oral exchanges, yet it preserves a written record, and allows optional off-line time for reflection between iterations, if needed, unlike spontaneous real-time dialogue. He believes that

> our talking heads and their interacting minds will be incomparably more fecund once those slow iterative cycles by which our knowledge had been created and accumulated in the Gutenberg era are restored to the speed of stone-age thought by skywriting in the PostGutenberg Galaxy (*Op. cit.*).

For one thing, I am very skeptical about the speed of stone-age thought being of the same order of magnitude as the information-age thought. Nor am I convinced that, in the course of evolution, thought itself has not undergone dramatic qualitative changes which disallow of the analogy between the stone-age and modern times. But even these considerations aside, one can hardly agree with the analogy drawn by Harnad between the PostGutenberg Galaxy and the oral tradition. In Dan Sperber's opinion, expressed in the discussion of Harnad's paper, the disanalogies between the two are overwhelming, and they have to do, at the cognitive level, with both short term and long term memory. As for short term memory, argues Sperber, seeing something written allows one to examine it in a way never available with an oral utterance, and makes genuine detailed criticism possible. The kind of "creative disagreement" Harnad talks about is narrowly tethered to short term memory in a genuine oral tradition; thus, the "quote/commenting" activity practiced in skywriting is just not a possibility (trivially, points out Sperber, a "quote/commenting" disagreement quickly turns into an "I didn't say this"/"yes you did" argument of little intellectual merit).

Turning to long term memory, Sperber reminds that, in an oral tradition, only things that are repeated again and again are reliably remembered: shared oral traditions are tethered to individual long term memory. Of course, there is internal differentiation within the group sharing an orally transmitted tradition, allowing a richer culture than could be held in one individual's mind, but this amounts to saying that the sharing is partial. Internet culture is just the opposite: not only is the amount of information that can be stored, and almost immediately accessed by every individual, unbound, but also, with research engines such as Google, we are, if anything, better at retrieving specific information from the web than from our own mind. So, concludes Sperber, far from heading "back to the oral tradition," in our technologically advanced use of writing we, in many essential respects, are getting even further away from it.

I could continue bringing up more and more controversial points raised in Stevan Harnad's paper, but I believe I have made it clear that his views on the cognitive relationship between oral language and writing are controversial because of a number of inconsistencies. What is important, however, is that the rationalist delusion, when writing is viewed as a code which is in one-to-one correspondence to spoken language which is also a code, affects cognition both on individual and social levels and is responsible for the spread of functional illiteracy in modern developed societies which leads to communicative dysfunction. Let us see what this is all about.

4.3. The Concept of Literacy and the 'Code' Model of Language

Until not so long ago the concept of literacy was trivially related to the ability to read and write, so a literate person was a person able to read and write and, as a consequence of these skills in a literate society, a person having an education. Literacy thus understood was (and still is) believed by many to be a crucial requirement for normal functioning of individuals in society. Education as a process purported to provide tools and techniques for cognitive development, usually referred to as 'knowledge acquisition', depends crucially on literacy necessary for the development of the powers of reasoning and judgment in the 'world on paper' (Olson 1994), and for preparing oneself or others intellectually for social life. It is, however, an indisputable fact that the ability to read and write of itself does not make one socially mature. What was attributed by the UNESCO projects of the 1950s and 1960s to literacy in terms of ability to empower was essentially imaginary (Olson and Torrance 2001) and, for example, the development of the power of reasoning turned out to depend not so much on literacy *per se* but on one's ability to establish meaningful relationships between verbal knowledge acquired in the process of education and one's life experiences. In other words, literacy is also "a set of prescriptions about using knowledge" (Cook-Gumperz 2006: 1). As Mr. Koïchiro Matsuura, Director-General of UNESCO, emphasized in his message on the occasion of International Literacy Day, 8 September 2006, "literacy is not merely a cognitive skill of reading, writing and arithmetic, for literacy helps in the acquisition of learning and life skills that, when strengthened by usage and application throughout people's lives, lead to forms of individual, community and societal development that are sustainable". These forms of sustainable development, in turn, depend on the ability to assign *meanings* to different phenomena which are then used to exert orientational effects on others via *language* as the environmental dynamic medium crucial for the existence of humans. Such orientational effects are *in-formation*, understood as integration of an organism into its environment as a result of which the organism becomes in-formed.

An actual realization of the entire education sequence depends on language as a specific cognitive domain of interactions aimed at in-forming young individuals in their physical and, above all, social environment. In the course of such interactions, or languaging, linguistic entities are generated which constitute a sign system whose philogenetic function is to mediate between generations what we are used to calling 'knowledge' as linguistically categorized experience relevant for the adaptive power of the species. In literate societies, the domain of linguistic interactions includes two sign systems, the spoken and written language, whose ontologies and epistemological purport are different. For that reason, despite the long established tradition, speech and writing cannot be viewed as two different manifestations of the same phenomenon.

The importance of literacy can be put down to the fact that life in a modern (especially post-industrial) society is to a large degree (pre)conditioned by the medium of written (graphic) signs which make up texts. Every little step we make in our mature lives makes us interact with texts either by reading and interpreting those designed to affect us and our behavior, or by creating (writing) texts designed to influence others. Texts are everywhere, they are omnipresent and omnipotent as they literally determine every aspect of human life. Even the progress of civilization itself, the development of industrial and post-industrial societies have been seriously affected by writing as one of the greatest inventions of humans (Hagége 1996).

In keeping with the traditional concept of literacy, a literate person is one capable of both 'extracting' knowledge from texts and creating his own texts to 'transfer' knowledge. In other words, texts are viewed as codes used to convey meanings, or thoughts, from one person (the writer) to another person (the reader). These encoded meanings (or 'mental content', 'abstract knowledge', 'complex propositions', etc.) are assigned to special entities known as '(natural language) signs' in order to be conveyed from the speaker's head into the listener's head via the medium of oral speech while writing is believed to be a system of graphic representations of these signs (such as words) — hence the so-called 'code model' of communication based on the 'conduit' metaphor (see, for example, Hinrichs 2006). On this traditional view, "language is in the mind", and linguistics as the scientific study of language is called upon to tell us important things about the architecture of the mind (Culicover 2005). In this framework, literacy is believed to empower individuals by allowing access to knowledge 'stored' in written texts, hence the concept of 'knowledge acquisition' as a process whereby a literate person becomes more 'knowledgeable' by reading texts and extracting knowledge contained in them.

Although traditional linguistics views natural language as a code, the concept of code is applicable only to conventional systems of graphic signs (such as letters of an alphabet) which symbolize ('represent', to use a conventional term) sounds and sound clusters characteristic of a particular spoken language. The number of speech sounds (sound types) in a given language (identified as 'phonemes') constitutes a closed set, and it is the socially sanctioned and prescribed correspondence between a particular letter and a particular sound (cluster) which makes the letters of an alphabet also a closed set and the relationship between the letters and the sound one of *coded equivalence.* However, as soon as a higher level of graphic representation is reached — that of words — the concept of code becomes inapplicable because of the open and indeterminate nature of the set it is supposed to define, i. e., the lexicon. Words of natural (spoken) language are signs interpretable in more than one way. These interpretations (meanings) form *fuzzy sets,* and words, therefore, do not (and cannot) qualify as units of a code. As emphasized by Ruben Budagov (2000: 131), "any code is a closed rigid system, while a living language is an open dynamic system with an immense potential". If letters are viewed as signs which stand for particular sounds, the relationship between the letter and the sound (that is, the 'meaning' of the letter) is relatively (on the historical time-scale) stable and unambiguous. What the letter-sign stands for cannot be changed voluntarily on the spur of the moment as it would undermine the integrity of the code and result in unrecognizability and uninterpretability of concatenations of the elements of the code (written words). But this is exactly what we observe in how the system of natural linguistic signs functions: there is no clear-cut stable correspondence between a word and what it stands for. This empirical fact lies at the root of the problem of linguistic meaning — just because written language is not the transcribed spoken language, it is an entirely new linguistic phenomenon, a cultural artifact, and as such it is external in regard to natural language.

4.4. Speech and Writing as Cognitive Domains

The role of writing in human society can hardly be overexaggerated because the use of graphic signs made it possible to overcome the limitations imposed on the cognitive potential of natural linguistic signs (words and sentences), that is, their potential to exert orientational effect across different time scales, becoming a major developmental factor in phylogeny. One of the obvious limitations imposed by the sound matter on the cognitive potential of acoustic linguistic signs is the limited time of their existence. On the one hand, sound waves travel through the medium along the centrifugal vector, so humans (typically) perceive acoustic signs as unidirectional short-lived auditory stimuli. On the other hand, human vocal cords are not some kind of *perpetuum mobile*: we get tired from long speeches, our voice may become hoarse or we may lose it altogether as when we get a bad cold; finally, a person may simply die 'taking' with him the knowledge which he possessed and which now cannot be accessed via his linguistic behavior as an informational medium.

Another limitation is the physiological threshold for auditory perception, that is why the distance between linguistically interacting humans in real time-space is of such importance. It was this factor that led to the invention of sign systems based on visual perceptibility of signs, such as various kinds of semaphores, for example. But as soon as the natural bioanthropogenic acoustic substance of the physical body of the sign was sacrificed for the sake of specific pragmatic objectives and replaced by the substance of a different kind, the linguistic sign ceased to be a sign of natural language. A system made of graphic signs, by enhancing one of its functional aspects (for example, its capacity to sustain developmental continuity of humans in phylogeny by mediating relevant experience across time and space between generations and communities), considerably undermines its other aspects, such as the intrinsic capacity of speech to be integrated with other kinds of activity which together make up oral communication whose complex dynamics have a greater orientational effect on individuals than interaction with written texts does. The emergence of writing defied the limited cognitive power of natural linguistic (acoustic) signs caused by their groundedness in specific physical and socio-cultural context. As soon as the ties between linguistic signs and their ontological source (individual speakers in their unique physical and bio-socio-cultural contexts) were severed, the evolution of human society entered a new cognitive stage, the stage of *scientific knowledge of the world*, for scientific knowledge is that which reaches beyond the domain of sheer phenomenology, and the functioning of natural language is, in large part, exactly phenomenological.

In the cognitive philosophy of language meaning is understood as a cognitive function whose argument is *experience*, it is constructed from interactions with physical, biological, social, and cultural aspects of the environment. Experience is specific relative states of neuronal activity caused by past interactions with the environment and retained in time to serve as representations of such interactions. By interacting with these representations, cognizing individuals enhance their adaptive power by avoiding direct physical interaction with potentially hazardous aspects of the environment. As meaning emerges from interactions with the world and is, therefore, of experiential nature, its structure and content in each particular case are determined by a history of a cognizer's interactions with his niche (*Umwelt*) which includes linguistic phenomena as well. Since the experience of two separate individuals cannot be identical in principle, structurally and content-wise it implies inevitable non-identity of meanings associated by these individuals with one and the same object/phenomenon. As argued by Bertrand Russell (1948: 4), "the meaning that the child comes to attach to the word is a product of his personal experience, and varies according to his circumstances and his sensorium".

When using language, we draw on linguistic signs acquired in the course of accumulating personal experience of the world, and we do it in the spatio-temporal, historical, cultural etc. context which also includes language itself. Different facets of this experience are projected, one way or the other, on the socially sanctioned interpretations of particular linguistic signs (their 'common' meanings), so for every individual languaging cognizer a particular linguistic sign is loaded with unique associations with various things and events borne of individual experience. But how, then, is understanding in communication possible at all?

The degree to which we can understand others directly depends on the degree to which our personal experience as language users overlaps with similar experience of others, that is, on our consensual domain of interactions as a domain of mutually defined behaviors. Languaging is behavior in a consensual domain, and understanding of linguistic signs is a cognitive process whereby the cognizer looks for common experiential ground in interpreting linguistic behavior of others. That is why in the biological theory of cognition language is viewed as connotational rather than denotational. But if language is not denotational, it simply cannot be defined as a code! The root of the problem with the code model lies in the fact that graphic representations of natural linguistic signs (physical words) constituted by code symbols (letters) standing for the structural elements of physical words (that is, sounds), are vitalistically explained as encodings of what they stand for (spoken words) and, by extension, of the spoken words' meanings. Such vitalistic explanation, however, cannot be but erroneous.

Live language use is linguistic behavior in a consensual domain. Adequate interpretation of such behavior depends on taking into account the physical context in which linguistic behavior is enacted, and the alphabetic code does not allow that by definition. Indeed, how can letters 'represent' the various prosodic features of speech, gestures, facial expressions, particular looks, etc. that are integrated in the complex dynamics of oral communication? But in a canonical situation of communication (situation of confrontation) all these so-called 'extralinguistic' factors often play a crucial role in the interpretation of linguistic signs. However, these shortcomings of the written language, paradoxically, turn into an advantage because, as a result of cutting off the 'extralinguistic' part in the graphic representations of natural linguistic signs, a new type of linguistic environment (cognitive domain of interactions) emerges, namely, *the domain of (written) texts*.

The ability of language users to interpret texts, or the orientational effect written texts can have on an interpreter's behavior, largely depend on the interpreter's experience of this type of linguistic environment projected on his life experience in general, or 'background knowledge'. This background knowledge derives in large part from interactions in the domain of texts whose function is to mediate knowledge as linguistically categorized experience of our species' interactions with the world. Thus, for the interpretative activity in the domain of texts to be successful, a large amount of various kinds of background knowledge is required, which orients the observer toward a particular interpretation (behavioral response). The increment of background knowledge is directly related to how a language user interacts in the domain of texts, and the context in which a graphic word is used — or, rather, the combined experience of such (highly variable) contexts that an interpreter possesses as a result of his interactions in the domain of written texts — affects interpretation in a decisive way. This is a very important point to be made about interpreting ('understanding') written texts.

However, in education viewed as a guided process of knowledge acquisition via language, the code model of language rules the roost. Because language is believed to be a code, language acquisition is viewed as getting the mastery of an inventory of linguistic signs (the lexicon) which possess certain meanings (allegedly attached to these signs and registered in dictionaries), and of rules for combining these signs (words) into meaningful sequences (the grammar). Taking stock of what is included in such inventory is accomplished by means of graphic representations (writing) viewed as the product of the use of a special code. Hence literacy is understood as the ability to perform a three-step operation of matching and identification:

(i) a particular letter (graphic code sign) is matched with a particular acoustic phenomenon (sound type, or phoneme),

(ii) a particular concatenation of letters (graphic word) is matched with a particular concatenation of sound types (spoken, or physical word),

(iii) the graphic word is identified with the spoken word.

This ability is, presumably, manifested in the acquisition of reading and writing skills, and a person possessing such skills is believed to be literate. Yet such view on literacy is totally unacceptable.

It is to state the obvious to argue that no one would think of a child who has learned to use a pencil to draw unsophisticated lines and figures on a piece of paper as an artist. Likewise, an older child who has learnt to produce all the sounds of the octave on the piano and match them correctly with particular graphic signs used in music notation, does not become a musician or a composer, although he can read the music score (is musically 'literate'). We won't say about such a child that he can *play* the piano. But if it is a child who has learnt to vocalize, more or less correctly, graphic representations of words, we are tempted to say, "He can read". Can he, really?

As an experiment, undergraduate students (Linguistics majors) in one of the Russian universities were offered to read and interpret a sentence I made up disguising it as an excerpt from a 19th century Russian novel. The sentence contained 19 words and was an account of the scanty belongings (totaling four items) of a person named Triffon (now deceased):

(7) *Podmochennaja shtuka bumazei, skladnoj derevjannyj arshin,*
Dampened piece flannellette G, folding wooden arshin[21]

[21] *Arshin* — old Russian unit of measure (~ 71 cm); *chugunnyj bolvan* — an iron heated on a stove; *shtoff* — a square short-necked bottle (1.2 *l*)

staryj chugunnyj bolvan I pochatyj shtoff lafita — vot vsjo,
old cast-iron block and begun Adj bottle Lafite G that all

chto ostalos' posle Triffona.
what remain PT after Triffon G

'A damaged piece of flannelette, a folding wooden arshin, an old cast-iron block, and an open bottle of Lafite — that was all that was left after Triffon.'

Not only did the testees fail to understand what was left after Triffon's death, which was to be expected as the things mentioned in the sentence have long gone out of use and, therefore, are not part of the younger generation's cognitive domain, but — and largely just because of this initial failure — they also failed to 'extract' larger chunks of knowledge 'contained' in the sentence, which were as follow:

(i) Triffon was a cheap garment maker,
(ii) he died in poverty,
(iii) he used to know better times.

Inferences (i) and (ii) follow from the description of his belongings (remarkably, some of the testees failed even to understand that Triffon was deceased), and (iii) is derived in part from the name of one item (an open bottle of Lafite) and in part from background knowledge which tells an educated interpreter that vodka in old Russia had always been very cheap (much cheaper than wine), and that if a poor garment maker spends what little money he has on more expensive alcohol, it means most probably a strong habit which grew from long practice at the time when Triffon could either afford red wine (less likely), or had access to it as when being a serf to a Russian nobility (more likely). Of course, there is also a whole specter of other inferences of a more hypothetical character one could make while interpreting the test sentence, but the point in case is that 'literate' people with a nearly completed 'higher education' could not make head or tail out of a simple sentence in their native tongue. As much as it is disturbing, it is also indicative of some anomaly in the cognitive development of young individuals whose cognitive powers modern education is called upon to enhance. This anomaly is the direct consequence of viewing writing as a code which represents spoken language which thus also is understood as a code — quite erroneously. Both spoken and written language are a kind of cognitive scaffolding which humans use to orient themselves and others in the cognitive domain of interactions, and the types of linguistic interactions involved in each case differ essentially from one another.

4.5. Types of Linguistic Interactions

Our perception and cognition of the world are rooted in interactions with the environment which is psychologically structured as follows: (i) personal space, (ii) home territory, (iii) public territory (Lyman and Scott 1966; Sommer 1969; Slane, Petruska and Cheyfitz 1981; Kravchenko 1992). As a structured domain of interactions public territory is subdivided into three sub-domains: educational, intracultural, and cross cultural. The first and most important stage in the process of language acquisition, when a young human organism actively in-forms itself in linguistic environment, occurs within personal space and home territory. It is this domain of linguistic interactions that moulds a person's initial standards of linguistic behavior; the adequacy of this behavior is continuously tested through direct or indirect appeal to authority (parents and family members). The adequacy of a child's linguistic behavior is assessed by relating it to the family members' personal values, priorities and idiosyncrasies whose nature and character, as well as public relevance, may vary considerably.

As a young individual grows up, his domain of interactions with the environment expands into the domain of public relations. Before a young human, however, functionally integrates into society and its culture as a relatively autonomous member, a long period in his developmental history (roughly between 11 to 25 years in economically developed countries) falls on the stage of educational interactions whereby, supposedly, he must complete the process of in-forming in the human habitat. During all this time, based on the already established standards of behavior in his consensual domain, an individual directs his adaptive efforts at eliciting encouragement and approval on the part of those whom society has sanctioned to do so, that is, all those professionally involved in education. They all have their personal values and preferences, and they all interact in multiple consensual domains pursuing essentially similar, though contending, goals — to adapt to the environment as best they can. A human individual, whose linguistic behavior continues to be influenced by the educational domain of interactions, is under the pressure to build his interpretative strategies taking into account numerous contextual parameters of his consensual domain which enable him to judge whether his own linguistic behavior has been 'approved' by those who have respective authority — and he uses the same model as in the domain of family/home interactions. As a result, when interacting in the domain of written texts he is apt to structure his linguistic behavior (interpretation of texts) with reference to how he interprets the current physical context which has nothing to do with the text to be interpreted by definition. The important consequence of this is that graphic signs which are very close to being code symbols thus making predictable interpretation possible on the part of different language uses regardless of their time and space (which, as it were, is *the* socially relevant function of writing), are perceived as natural linguistic signs whose interpretation directly depends on the time, place and other circumstances in which they are used. This lays the ground for the functional failure of linguistic behavior (both reading and writing) in the domain of written texts, or *communicative dysfunction.*

A trivial example of communicative dysfunction on the part of the writer would be a notice attached to an office door, such as:

(8) *Be back in 20 minutes.*

By creating this text, that is, by entering the domain of specific linguistic interactions, its author identified graphic representations of linguistic signs with signs of natural language as a consensual domain of interactions. Thus the author of the notice stripped his linguistic behavior of its ability to exert orientational influence for the simple reason that *there is no consensual domain* in this case. As a result, the interpreter (the reader of the notice) cannot establish a common point of reference and may only guess whether he will have to wait 5, 10, or all of the 20 minutes (granted that the author of the notice is true to his word). Now, this is just one possible interpretation of the situation based on the assumption that the person who wrote the notice is not cognitively/linguistically sophisticated. Another possibility is that whoever wrote it realized fully well what he was doing, which was providing himself with a cover for his prolonged absence from the workplace in case his immediate supervisor happened to drop by during that time.

On the part of the reader, failure to functionally interact with the domain of written texts must also be attributed to the interpretation of written (codified) signs as if they were natural linguistic signs. Let us take, as an example, such Russian expression as *udostoverenije lichnosti* (*ID* 'something that identifies one') and the way it is interpreted in the domain of public interactions. This expression when used in the indicated sense usually refers to a document issued by a government or public office which has been authorized to do so by society; an ID contains a photograph of its owner and his name, plus the name of the issuing authority and the name of the document itself. Examples of IDs would be a passport (national, foreign, diplomatic), a military service card, a driver license, a student card, a reader's card, a work pass, etc. To establish one's ID is to authoritatively (that is, by the authority delegated to a particular office or institution by society) verify that the person whose photograph is on the ID is one of the name stated on this ID. In real life, however, the situation is somewhat peculiar as, counter to the codified meaning of the written expression *udostoverenije lichnosti*, government officials and all sorts of bureaucrats (and it is they who demand citizens to provide IDs) accept as such only two documents: the national passport for civilians and the military service card for those in service. If, for example, you want to make a withdrawal from your bank account and use your driver license as an ID, the teller will tell you that *it is not an ID*. The same is true about purchasing railroad and airline tickets, filing employment applications, checking in at hotels and so forth — in all these cases the only types of ID that will facilitate the activity in question will be either the passport or the military service card. Yet even one's passport will not be accepted as a valid ID at the bank if it is a foreign passport rather than national. Although the official customer service regulations issued by the Central Bank of Russia stipulate that an acceptable ID is a valid Russian passport without specifying whether it should be national, foreign, or diplomatic, bureaucrats at the banks interpret it very narrowly as referring only to the national passport. How can this weird situation be explained?

The explanation is rather simple. A bureaucrat as the typical rank-and-file interpreter of the expression *udostoverenije lichnosti* is driven by his desire to use, as a cognitive scaffolding, the authority of another interpreter in the consensual domain whose linguistic behavior, because of specific circumstances, produces maximal orientational effect on the bureaucrat's linguistic behavior. In other words, the bureaucrat is guided by the opinion of his supervisor, or *the boss*. Unfortunately, interpretative abilities (literacy skills) of the higher-ups in Russia in regard to written texts often leave much to be desired because of the lack of sufficient cognitive experience borne of life-time interactions in the domain of written texts. In other words, their developmental history during the maturation period did not feature *extensive reading and writing* as activities *crucial* for human cognitive development in the information era. It is something only to be expected from individuals aspiring to 'make it to the top', for it is interactions in the domain of natural linguistic activity (being oriented by others, such as the higher-ups, and orienting others by return) that acquires to them paramount importance in adapting to the immediate environment. They simply underrate the cognitive importance of the domain of written texts because to them it is *people*, those *in command* who mould their immediate environment, not some dead artifacts. The jocular Russian:

(9) *Ty nachal'nik — ja durak, ja nachal'nik — ty durak*

You boss I fool, I boss you fool

'You're the boss — I'm the fool, I'm the boss — you're the fool'

reflects a very real state of affairs in the domain of interactions with written texts onto which the praxis of natural linguistic interactions is extrapolated. From here there is but one step to understanding the innermost wisdom of the saying which became especially popular among more enlightened Russians in the last decades of the Soviet era:

(10) *Rossija — strana durakov.*

Russia land fools G

'Russia is a land of fools'

Despite the fact that many Russians agree with this self-deprecatory statement, it is virtually impossible to find people who would believe or admit themselves to be fools. A fool is always someone else, it is always not-I, it is someone whose (linguistic) behavior in the domain of public interactions is evaluated as non-sensical from the point of view of the interpreting observer. But what are the criteria for 'sense' in each particular case? Do we mean something intrinsically characteristic of an individual, or is it something socially determined? Is it the meaning we ascribe to natural communicative or codified written linguistic behavior? The truth of the matter is that we make judgments about one based on our interpretation of the other, and this, quite naturally, leads to an epistemic paradox:

(11) *Russia is a land of fools. I live in Russia. I am not a fool.*

This paradox is brought about by the impossibility (or our unwillingness) to take the stance of an impartial observer of our own domain of linguistic interactions, by our readiness to judge others and our unpreparedness to judge ourselves. An individual incapable of rising above this incongruity in his cognitive development cannot hope to fit functionally in the intra-cultural domain of interactions in a society of which he is, at least nominally, a constitutive member.

4.6. The Bio-Socio-Cognitive Function of Speech and Writing

The concept of 'functional fit' in the intra-cultural domain of interactions is essentially controversial inasmuch as there is basic dialectic contradiction between the goals by which an individual on the one hand, and society on the other, are driven to particular behavioral patterns in their adaptive interactions with one another and the world. For an individual, functional fit in the intra-cultural domain of interactions means choosing a behavioral strategy that would maximize his autonomy as a cognizer in the consensual domain while minimizing the restrictive influence of the (social) environment. For a human community (society) as a living system, however, it is the other way around: it seeks to enhance its adaptive potential essential for the well-being of the species by minimizing the cognitive autonomy of its individual members. Thus, the intra-cultural domain of interactions (as well as the cross-cultural domain of interactions on a larger scale) is the flux of human cognitive activity evolutionarily driven by the necessity to reach dynamic equilibrium between the parts and the whole. So, what is the role played by individual linguistic interactions in sustaining conditions which facilitate the normal functioning of modern society? And what are these conditions?

Juxtaposing language and society, we should not forget about the ontological duality of language and the essential phenomenological difference between spoken and written language. To use the 'organism' metaphor in regard to human society, spoken language may be viewed as the 'nervous system', and written language as its central part which coordinates normal functioning of the different parts of the organism *as a whole*. First, signals from the central nervous system, being a neurological outcome of the 'interpretation' of the states of different parts of the organism as a functional unity, possess greater adaptive relevance over signals generated at the peripheral level of the vegetative nervous system. Second, the greater the complexity of an organism as a functional system, the greater the complexity of its central nervous system (the brain) which, in the case of higher vertebrates, becomes the 'soft spot' in terms of the organism's vulnerability and its normal functioning: consequences of physical damage to the brain by far outweigh comparable (in terms of body mass) damage to other organs. Third, functional 'glitches' and disturbances at the vegetative level may interfere with, or interrupt, the adequate (cognitively 'correct') interpretation of signals from the central nervous system. As a result, a particular organ may not be able to perform its function within the system and the organism's overall adaptive ability may decline.

Similar considerations apply to the relationship between spoken and written language as sign systems for regulating the functioning of human society as a living system. Every single individual speaking a language common to a given community is a necessary constitutive member of this community because individuals and community are mutually co-determined in language and through language. Linguistic communication and the establishment of a cognitive domain of linguistic interactions is a necessary prerequisite for the formation and development both of community and its individual members, therefore the primary function of language in regard to society is socio-generative. Within this socio-generative function different hierarchical levels of linguistic interactions should be distinguished; these reflect the hierarchical structure of the environment mentioned in Section 4. The domain of texts as a domain of linguistic interactions is ontologically secondary in regard to the (natural) linguistic consensual domain at the level of an individual, but it is primary at the level of society as a structure organized system — just because individuals and society are mutually co-determined. Consequently, the normal functioning of an individual in society which, in turn, pre-conditions the normal functioning of society itself, presupposes the primacy of orientational effects of the domain of written texts over the orientational effects of (natural) linguistic consensual domains. Different phenomenologies of the domain of written texts and the domain of spoken language call for different systems of interpretations based on different points of reference ('coordinate grid'). The optimal functioning of an individual in society thus depends on the degree to which his skills for interpretation in these different coordinate grids have been developed. Undeveloped skills for interpretation of orientational effects generated by interactions in the domain of written texts, as well as lack of skills for projecting the results of such interpretation onto the consensual domain of natural linguistic interactions, lie at the root of the phenomenon of *functional illiteracy*. In other words, functional illiteracy is the inability of an individual to modify his linguistic behavior by taking into account the orientational effects of interactions in the domain of written texts.

4.7. Summary

The above considerations by no means exhaust the range of issues related to the language/cognition problem. The functional relationship between the spoken and written language as ontologically different sign systems, and their role in individual and public cognition are only beginning to be brought under scrutiny. However, one important conclusion can be made even now. The recent but alarmingly growing trend in contemporary society for the scale and frequency of interactions of individuals in the domain of written texts to shrink, or the fact that people *read less and less*, is already causing functional disturbances which give rise to grave concerns and misgivings. If this tendency is not curbed in the near future, there is high probability that functional illiteracy on the part of individual cognizing subjects is going to spread like a malign tumor over the entire social 'organism'. It is time that human society undertook a massive coordinated effort to revise the very principles underlying the concepts of literacy and education (cf. Resnick 1987; Brady 1997; Žarnić 2002; Smith 2003) so as to enable young individuals to maximally develop their adaptive powers to respond to the ever more sophisticated orientational effects of human society of the post-industrial era. This is where linguistic education and a revised concept of literacy come to the fore, recognizing the important role writing plays in human society. As stressed by Mr. Koïchiro Matsuura, "in order to have an enduring impact, literacy needs to be nurtured and supported through the availability of books, newspapers, magazines, computers, and other modalities of written communication", and to promote an environment supportive of functional literacy is, according to the International Plan of Action for the United Nations Literacy Decade, to establish *a culture of reading in schools and communities.* Sustained culture of reading and writing is essential for the cognitive development of human individuals and for the philogenetic development of the human species as a whole as it allows to avoid cognitive/communicative dysfunction in the intra- and cross-cultural domains of interaction. However, a culture of reading and writing may be sustained only by effective interactions of individuals with written texts as cultural artifacts; interpretation of texts by readers must be based, more or less, on the same cognitive strategies as those used by writers who create texts following certain cognitive strategies. This is where the concept of language grammar comes in — to be discussed in Part Two.

PART TWO

THE LANGUAGING OBSERVER

CHAPTER 5

GRAMMAR AND COGNITION

5.1. Science and Common Sense

As, I hope, has become obvious from the discussion in Part One, a lot of what orthodox linguistic thought holds dear as 'truths' about language is, in effect, a skewed picture of a particular phenomenon reflected in the distorting mirror of folk theories bases on what we often call 'common sense'. Although some believe that mental activities are best understood in lay terms (Harris 2004), many would agree with David Spurret's (2004) expostulation:

> Either we think that science can tell us that we're wrong with how we think things are with us . . . even to the extent of showing our common sense, or vulgar, self-conception to be deeply mistaken, or common sense is holding some kind of trump so it always beats science, or even that it never has to pay any attention to science (p. 497).

We can, of course, persist in thinking that written language represents spoken language, that language is denotational and meanings are contained in words, or that language is in the mind (a mental organ) and consists of a mental lexicon and a set of grammatical rules (linguistic competence) used to express and convey ideas from one head to another via the channel of verbal communication (the conduit metaphor). At least, we would be justified in doing so by the fact that this is what common sense tells us — and anything it does tell us is in *linguistic* form, because, from the point of view of the biology of language, we human beings happen in language, and we happen in language as the kind of living systems we are. But progress in science is not possible unless someone begins to think 'unthinkable' thoughts. And any thought, as the history of science tells us, is 'unthinkable' as long as it goes counter to the views and common sense intuitions shared by the majority. Fortunately, the progress of science over the millennia of human history has reconciled (at least some of) us with the obvious incompatibility between the naïve theory of the world 'fossilized' in the language we use, and what science tells us about it.

On a historical scale, not so long ago the conception of the world in different cultures was very different from what children are being taught in schools today in the course of their introduction to natural science. Take, for example, the pre-scientific concept of the universe held by the Hebrews and their pagan neighbors. In the world of the Hebrews, the earth was a platform resting on columns and surrounded by waters, and the sky was a firmament resembling an overturned bowl supported by the columns of mountains. In other cultures, the flat earth rested on whales or turtles, or what have you. The important thing about such conceptualizations was that the earth *had* to be flat, — otherwise, common sense told prescientific thinkers, people and things and everything else would be bound to tumble off its surface. The pre-Copernican geocentric model of the universe was believed to be true because it was also built on common sense as revealed through reasoning done in the language we speak, yet it was eventually rejected by science, although the process was not quite that fast or easy or peaceful. The reason why transition from one paradigm to another should be difficult is quite simple — as noted by Bernard Barber (1961), scientists are often intolerant of new theories invented by others, particularly so when such a new theory lies outside the currently accepted and practiced paradigm.

One could continue with examples of 'scientific' views dominated by common sense. We could mention here the firm belief that apparatuses heavier than air could not fly, or discuss the effect of Albert Einstein's Theory of Relativity on modern science, or turn to the stunning discoveries made by quantum physics about the nature of matter. All these make clear one point — that we should not forget that at any particular time in its history science is, essentially, a set of beliefs based in tradition which itself is rooted in the institutionalized praxis of the previous generations of scientists. As observed by Thomas Kuhn (1962: 11), "men whose research is based on shared paradigms are committed to the same rules and standards for scientific practice". And much of linguistic 'science' practiced today is a set of beliefs, sanctioned by long tradition, about what it is to 'do linguistics' (the infamous 'language myths'). As observed by Don Ross (in press), the domain of linguistics

> is both the most transparent site of traditional errors about human peculiarity, and — for much the same reason — provides an ideal perspective from which to diagnose and correct those errors (p. 2).

The concept of grammar in orthodox linguistics as it is used by linguists committed to the same rules and standards for scientific practice based on common sense, is a good example of how erroneous science can be.

5.2. Approaching the Concept of Grammar

5.2.1. What Does the Term "Grammar" Refer To?

The written language bias in linguistics could not but affect how the object of study of linguistic science was conceptualized and legitimatized. Preoccupation of linguists with concatenations of graphic symbols (second-order cultural artifacts) believed to represent natural linguistic phenomena (first-order communicative activity) resulted in virtual substitution of the study of natural language, which is a dimension of complex integrated behavior of humans in a consensual domain of interactions, by the study of grammar as the art of writing (that is, the art of creating second-order cultural constructs). As a consequence, to many the study of language today means simply the study of grammar, and the term "grammar" itself has come to be used almost as a synonym to "language" (as in *generative grammar, functional grammar, communicative grammar, cognitive grammar,* etc.). Of course, if we speak of *written language*, then the use of the term "grammar" in reference to the principles and mechanisms involved in the organization of a piece of writing causes no objection. But we must not overlook the fact that writing, not being a representation of natural linguistic behavior (oral communication) by definition, is subject to rules and norms essentially different from the 'rules' and 'norms' of spoken language. I use single commas here to emphasize the sheer arbitrariness of these concepts when applied to spoken language. Indeed, if there be rules that govern natural linguistic behavior, they could not possibly be understood and formulated in separation from the cognitive dynamics of communicative behavior. This means that there is no such thing as 'purely linguistic data'. But how, then, are we able to single out and identify certain regular patterns in the communicative behavior of humans which we describe in terms of grammatical (i. e., regular and systematic) phenomena? Are we justified at all in our use of the word *grammar* when we focus on these patterns? What are these patterns based in?

The answer to the first question is *recursion.* In his *Biology of Language: The Epistemology of Reality* Maturana (1978) emphasizes that regularities in behavior pertain to the domain in which the behavior is described by the observer, not to the underlying physiology:

> Therefore, the describable regularities of the linguistic behavior of the members of a consensual domain do not necessarily reflect an identity of the underlying physiologies that generate the linguistic behavior of the different members. Only if the original structures of the consenting organisms had been isomorphic could some isomorphism be expected in the physiology of similarly behaving organisms that participate in a consensual domain. Such a coincidence, however, would be a matter of historical contingency, not of structural necessity (pp. 51-52).

According to the internalist account of cognition and language discussed in Chapter 2, linguistic behavior is what cognizing humans do as a result of the processes that take place in the brain, that is, it is defined by the structure and physiology of the organism. One of the main efforts of the generative enterprise — no wonder — has been to identify the specific part of such structure or physiology responsible for language, the so-called 'language organ' (among the recent contributions one could mention Jenkins (2000) and Anderson and Lightfoot (2002)). However, because regularities in linguistic behavior pertain to the domain in which the behavior is described by the observer, or to the relational domain of interactions in which, as Maturana points out, the behavior of an organism is defined under the conditions in which the organism realizes its autopoiesis, they are descriptions of the original structure of the organism at each interaction and are not determined by the nature of the selecting behavior:

> [T]he regularities or rules that an observer can describe in the performance of any particular behavior, whether it is courtship, hunting, or speaking by the different organisms that enact it, do not reveal homorphisms in the underlying physiologies (Maturana 1978: 51).

Or, simply, the fact that snakes and hedgehogs hunt for mice does not mean that their organisms have isomorphic structure. And this is precisely what the Turing Test was meant to achieve: to identify, through observing a particular 'human-like' behavior of a robotic system, the underlying structure (surely it would be too much to speak of a physiology in the case of a robot) 'responsible' for this behavior, and then look for a similar structure in a human organism. It is not surprising that first generation cognitive scientists have not been able to obtain the results they expected.

Still, since the entire concept of generative grammar is based on recursion of syntactic structures, and in the biological theory of cognition linguistic behavior is a kind of recursive behavior, it might appear that generativism does take a genuinely biological stance in approaching language, its nature and origin. But this is only superficially so. The following, necessarily lengthy, quotation from Maturana (1978) will help understand why:

> Every kind of behavior is realized through operations that may or may not be applied recursively. If recursion is possible in a particular kind of behavior and if it leads to cases of behavior of the same kind, then a closed generative domain of behavior is produced. There are many examples: human dance is one, human language, another. What is peculiar about a language, however, is that this recursion takes place through the behavior of organisms in a consensual domain. In this context, the superficial syntactic structure or grammar of a given natural language can only be a description of the regularities in the concatenation of the elements of the consensual behavior. In principle, this superficial syntax can be any, because its determination is contingent on the history of consensual coupling, and is not a necessary result of any necessary physiology. Conversely, the 'universal grammar' of which linguists speak as the necessary set of underlying rules common to all human natural languages can refer only to the universality of the process of recursive structural coupling that takes place in humans through the recursive application of the components of a consensual domain without the consensual domain. The de-

> termination of this capacity for recursive structural coupling is not consensual; it is structural and depends entirely on the operation of the nervous system as a closed neuronal network. Furthermore, this capacity for recursive structural coupling is at work both in spoken and in sign languages of human beings and in the sign and token linguistic domains established with chimpanzees <...>. Thus, the structure required for a universal grammar understood as a capacity for a recursive structural coupling in the operation of the nervous system is not exclusively human. The contingencies of evolution that lead in man to the establishment of spoken language, however, are peculiarly human (p. 52).

So, if recursion leads to cases of behavior of the same kind, a closed generative domain of behavior is produced. It is generative in the sense that a particular regularity in the behavior of one organism triggers a similarly patterned behavior of another organism and so on in recursive order. In dancing, for example, a particular move made by one partner 'causes' (triggers) a particular move of the other partner which, in its turn, leads to the next move of the first partner, etc. What makes language different from dancing, however, is that this recursion 'takes place through the behavior of organisms in a consensual domain', that is, the domain of interlocked conducts that results from *ontogenetic* reciprocal structural coupling between structurally plastic organisms. Linguistic behavior is not a kind of patterned behavior where each particular pattern (like a move in a dance) may be explicitly described with reference to the moves immediately preceding or following without any reference to the developmental history of the dancers. The dancers may be young or old, professional or amateur, in good health or ailing, but these factors have nothing to do with the actual 'structure' of the dance: if dancer A makes move X, dancer B makes move Y, and if dancer B makes move Y, dancer A makes move Z. The observed regularities in dancing are not instances of conduct resulting from the developmental history of the dancers. It is precisely this 'generative' model that Chomsky proposed, and his followers accepted and acclaimed, for the explanation of language. Viewing written texts as representations that stand in one-to-one correspondence with acts of verbal communication and erroneously identifying graphic disembodied symbols with actual linguistic behavior, which is holistically integrated with a wide scope of cognitive processes distributed over space and time and characterized by complex dynamics, adepts of generative grammar offer a model of linguistic behavior analogous to dancing. And because dancing a dance correctly is possible only if the dancers know the whole structure of the dance (its 'syntax'), which surely cannot be said about young children 'acquiring' language (the infamous 'poverty-of-the-stimulus argument'), it is concluded that humans are born with a 'language organ' responsible for their linguistic 'competence'. And of course nothing could be farther from reality.

Languaging is generative behavior in a consensual domain; it is, therefore, a description of the structure of the organism at the moment such behavior is enacted, while the structure of the organism itself is the result of a history of reciprocal structural coupling between organisms. A human baby, through successive stages of its development, is structurally coupled with other organisms such as the caregiver, for example, whose linguistic behavior plays an important role in the young organism's 'tuning' to the environment as a relational domain of interactions. Because language extends the human sensorium from which it cannot be separated either for ontological or epistemological reasons, and because human organisms as living systems are structure determined, linguistic interactions that constitute a dimension of the cognitive domain of an organism become the relational domain in which *humans exist as unities of interactions*.

Although common sense tells us that the regularities we observe in the linguistic behavior of others (i. e., recursion) are 'independent' of the languaging subjects as organisms with unique individual developmental histories, to persist in this line of thought is to play common sense as the unbeatable trump again. To begin with, there is much less regularity in natural linguistic behavior (spoken communication) than in graphic artifacts we call 'written texts'. One could easily see the indisputable truth of this claim by taking a quick look at any of the many available corpora of 'spoken language' (which are, still, only *transcripts* of actual discourse and do not represent the physical context in which transcribed linguistic behavior was enacted). Doesn't this mean that we should, probably, speak of *two* types of grammar for any single language with a writing tradition? Because if grammar is the art of writing, how can we justify our use of the word *grammar* when we focus on particular patterns in the natural linguistic behavior of humans, particularly if we don't believe that writing stands in one-to-one correspondence to oral communication?

5.2.2. How Systemic Is Grammar?

Viewed as 'the system by which the words and morphemes of a language are organized into larger units, particularly into sentences' (Trask 1993: 121), grammar has traditionally been a much harder object for analysis in linguistics than lexicon, which fact is well documented in a vast array of conceptual frameworks within which the grammar of a particular language may be described from perspectives so different as to almost make such descriptions incompatible (a ready example would be generative grammar as opposed to cognitive grammar). One possible explanation may be that in order to describe something as a system we must have a more or less clear understanding of (i) what are the characteristics of a system in general and how applicable they are to grammar, and (ii) if grammar is, indeed, a system, what *ultimate purpose* this system serves. Various co-existing radically different theories of grammar suggest that such understanding has not been attained.

That grammar is a system has been tacitly assumed by orthodox linguistics and accepted as a truism not worth serious discussion. But what is a system? If we look in a dictionary such as *Webster's Encyclopedic Unabridged Dictionary of the English Language* (compiled, mind you, by linguists), we will learn that a system, basically, is either an assemblage of (correlated) things/parts/facts forming a complex or unitary whole (*a mountain system; a system of shorthand characters; a living system*) or a coordinated body of methods/procedures (*a system of government; a system of classification*). As an assemblage of correlated things, grammar is usually described in terms of morphological and syntactic categories, and as a body of procedures it is described as a set of rules. As a class of formal objects, a grammatical category is a set of correlated linguistic items, such as words or phrases, the semantic distinctions within which are expressed by variations in the form. Thus, the English *walk, walks, walked* are an instance of the morphological category 'tense', and the Russian *letet', letat', proletet',* etc., which are formal variations of the lexical verb FLY, are an instance of the category of aspect. Grammatical categories have meaning which differs from lexical meaning; while the latter mostly bears on specific discrete particulars singled out in the perceived world — such as things, processes, states, features, etc., — the former reflects the relations between/among such particulars. For example, the meaning of the grammatical category of tense is traditionally defined as temporal relations between events identified with respect to a certain point of reference, such as the speaker or the moment of speech.

As a set of rules, grammar allows us to speak about the world around us and about our life in this world. This is achieved by combining lexical and grammatical meanings when we arrange specific forms of particular lexical items into specific strings or sequences (phrases, sentences, texts) making an obligatory choice between two or more forms distinguished within a particular category. This obligatory choice is a specifying feature of any grammatical category. For example, if I use a verb to refer to a real event or process, I must use one of the available tenses, as in the English *I walk to my office* as opposed to *I walked to my office,* but I can't possibly say **I walker to my office* because the lexical item *walker,* formally (and, therefore, semantically), does not belong to the category of tense. It is worthy of note, however, that were I to answer someone's question about how I get to work with: **I walker to my office,* it wouldn't cause a problem on the part of that person in understanding my meaning. Thus the actual choice of a grammatical form is not, strictly speaking, unconditional and appears to depend on what generativists call 'competence' and what I call *sufficient experience of linguistic behavior in a consensual domain.* And of course, 'sufficient' is a contingent notion — that is why grammaticality judgments made by different linguists discussing similar concatenations of linguistic signs often vary.

Once a complete inventory of categories has been made and the set of rules for combining different forms identified, we have the 'competence' necessary to generate any number of grammatical (that is, "correct") sentences. At least, this is what generative grammar tells us, and this is what thousands of Chomsky's followers world-wide have tried to accomplish over the past fifty years. And they have failed miserably, because they haven't been able either to draw a complete inventory of categories or to come up with a finite set of rules. Nor, for that matter, can other competing frameworks, such as functionalism, boast of considerable achievements in the study of grammar if by such study one understands the explication of a finite set of rules used to put words and phrases together to express meanings. And to make this trivial observation is to beg the question, "Why?".

While I don't presume to have a definitive answer to this question, I will make a tentative suggestion that the concept of grammar itself as applied to language has been misunderstood, and this misunderstanding goes back in time to the first Greek grammars which were, primarily, guides to good (that is, coherent and comprehensible) *writing*. A person who knew letters (Gk. *gramma* 'letter', *grammatikos* 'one who knows letters') was a person able to write so that later he himself or other persons who knew letters could understand what had been written. And this task was very demanding, because the mechanisms involved in understanding writing were essentially different from those involved in understanding oral speech (some of these mechanisms were discussed in Chapter 4). Written language does not represent oral speech as a specific kind of cognitive interactions with the social environment; while speech as dynamic behavior distributed in real time is *embodied*, and its inherent dynamics must be subject to on-line monitoring and analysis by both communicating parties, written language is *disembodied* with respect to the human agent. Linguistic behavior — or what is usually referred to as 'natural language' — is not an autonomous activity ontologically independent of other kinds of human activity, it is integrated with the complex cognitive/behavioral dynamics of the human agent and is interpreted by the observer as such, that is, as behavior in the observed physical context (relational domain of interactions between the communicating parties) in real time, when every little thing counts. Facial mimics and bodily gestures, variations in the speed, cadence, pitch, timbre and tone of speech, posture and gaze direction, what one of the communicating parties could observe about the protagonist in the situation preceding verbal exchange and inferences based on such observation, etc., etc. — all these are taken into consideration just as human agents engage in verbal communication. And precisely because the function of language is *not* to convey meanings but to *exert orientational effects* on others in the course of real-time interactions in a consensual domain, the structure of our normal speech in everyday discourse is far from conforming to the standards set by so-called grammar rules. Moreover, should one attempt to abide by the rules set by some *Guide to Effective Grammar* in a mundane situation such as verbal exchange among family members at breakfast or between two lovers in their moments of intimacy, it would be interpreted as a sign of basic social inadequacy if not mental disorder.

Written language is quite different in this respect. Just because it is not face-to-face communication in real time, none of the contextual information that serves as a scaffolding in constructing meaning in oral discourse is available. This disembodiment of written texts makes it imperative that the missing contextual information be compensated for, and since such compensation is possible only by drawing on the resources provided by written language and *in* written language, this is where grammar as the 'correct use of letters' steps in. As it is, our conception of grammar as 'a system for organizing words into sentences' is heavily biased by our preoccupation with texts as cultural artifacts. Modern human culture is, basically, text-oriented, and we as humans have to live in, and adapt to, a specific kind of environment constituted by texts which have become one of the most important and ubiquitous products of our biologically and socially determined behavior. Although it is sometimes claimed that grammar is not all that relevant in the organization of meaningful verbal communication (A. Gross 2005), it would be inconceivable to imagine how texts of different kinds could be made meaningful and coherent without the backbone of grammar which, as it were, molds the flesh of the lexicon it can support in building the body of a particular text. Thus the importance of grammar as a certain, although at times fuzzy, set of regularities in the use of written language responsible for the organization of texts must not be underestimated.

Neither must it be overestimated when we are concerned with spoken language. For one thing, in trivial acts of verbal communication such as curt exchanges among the family members throughout the daily routine, paying at the counter for purchased goods or services, swapping judgments with neighbors while watching a football game and the like, we seldom speak in complete sentences. We just *don't need* to do so, and as a rule it doesn't affect our understanding. While orthodox linguistics views language as denotational, and speaking is regarded as the expression of 'mental content' in the form of propositions constituted by particular denotants, in the biological theory of cognition language is *orientational activity in a consensual domain.* And orienting just isn't denoting. This is why, among other things, cases of what linguists label as 'ambiguous use' are seldom encountered in normal oral discourse (see Chapter 2 for discussion). The so-called 'ambiguous uses' when the meaning of a sentence may, allegedly, be interpreted in at least two different ways are, in fact, disambiguated by the numerous factors and parameters of the physical context in which they occur and which are not accounted for by traditional grammatical theory. Considerations of this kind (and of course, there are many more I haven't mentioned) bring up the question of the *degree* to which grammar may be characterized as a system in either of the senses, that is, as an assemblage of things (i. e., grammatical phenomena) and as a coordinated body of procedures.

Let us consider whether grammar as an assemblage of things is a complex or unitary whole. On the one hand, the way grammar has been treated (analyzed and described) in orthodox linguistics suggests that it is viewed as a unitary whole, otherwise titles such as *A Contemporary Grammar of English, A University Grammar of English* and the like we find on the covers of numerous books on grammar would carry no particular meaning. But it is also a well-known fact that no single grammar gives a complete and exhaustive account of all the grammatical categories and their meanings — not because grammar writers don't do their job of identifying and explaining the categories which are 'out there', but because the very nature and number of such categories depend on a particular theoretical framework endorsed by a particular grammarian. Even when we are told that some grammatical categories occur most frequently across languages while others are relatively rare, or that language X possesses categories not found in language Y, we should not overestimate the empirical validity of such claims because, being expressed in a particular language, they reflect a *particular world view* characteristic of a particular culture; therefore, they may not be regarded as universal truths by definition.

When we speak of something as a unity what we really mean is 'oneness'. For example, a living system such as a human organism is a unity since we can always tell whether we are looking at one or more than one organisms (we do not consider borderline cases such as pregnant females or Siamese twins). It doesn't usually take any special effort to tell where one organism ends and another begins. Yet it is not so with grammar. Various theoretical frameworks notoriously lack clear-cut definitions of what may be labeled as "grammatical phenomena proper". Actually, the realization that it is virtually impossible to draw a divide between 'purely grammatical' and 'purely lexical' phenomena within any given language facilitated the *functional/cognitive* approach to grammar (Halliday 1985; Langacker 1987; 1991; Geeraerts 1988; Givón 1995; Heine 1997) with spin-offs such as *construction grammar* (Goldberg 1995; 2005; Goldberg, Casenhiser and Sethuraman 2004; Shibatani and Thompson 1996; Cameron-Faulkner, Lieven and Tomasello 2003; Fried and Boas 2005) and *experience-based grammar* (Bod 1998; Tomasello 2000; 2003; 2005). As many linguists concede today, the orthodox view of grammar as a prescribed set of rules used to combine linguistic forms to express propositions about the 'objective world' may not be accepted as adequate anymore — even though for the single fact that such view cannot account for linguistic variation, when one and the same aspect of the 'objective world' may be 'expressed' by different language users in a number of different ways. The orthodox linguist will argue that instances of linguistic variation are simply surface manifestations of some invariant (the 'deep structure') which belongs to the realm of the mental — forgetting that there is no abstract mind (even if we do believe that mind is a thing), there are only individual (although dialogical) minds with no two minds identical to one another. Because, to the traditionalist, the function of language is to communicate truths about an objective world, and this objective world appears as the same to all speakers not only of the same language, but of all the different languages either living or dead, there may be no different truths about the same objective thing, only different 'surface structures'. Thus the notion of Universal Grammar is invoked as "those grammatical properties which hold for the grammars of all existing and possible natural languages" (Trask 1993: 293). But what is a grammatical property?

If grammar is the system by which the words and morphemes of a language are organized into sentences, then "grammatical property" refers to some specific part of this system viewed both as a thing and as a process. For example, particular morphemes or constructions in different languages said to express similar grammatical meanings, such as "thingness", "plurality", "agency", "possession", "temporality", "aspectuality" — to name but a few — would be "universal" grammatical properties and thus constitute Universal Grammar as a body of abstract concepts common to all natural languages, which are sometimes called "semantic universals" (Wierzbicka 1996). However, as has been repeatedly shown in the literature, there is rarely one-to-one correspondence between the allegedly similar grammatical meanings in different languages. Suffice it to mention such trivial grammatical meaning as "plurality" expressed in the category of number —

> [T]he grammatical category, most often associated with nouns and pronouns, whose primary correlation is with the number of distinguishable entities. English has a simple two-way number contrast between singular and plural <...>, but some languages exhibit more elaborate number systems involving dual, trial and paucal forms as well as singular and plural. Except perhaps in pronoun systems, number is not universally present in languages; Chinese and Japanese are two examples of languages in which number contrasts are generally absent (Trask 1993: 192).

Even an authoritative dictionary such as the one quoted above can do little more than suggest a conceptually controversial definition of what appears to be a rather trivial grammatical category — as compared, for example, with the much more problematic categories of tense or aspect. On the one hand, this category is 'most often associated with nouns and pronouns', yet on the other hand, 'number is not universally present in languages except perhaps in pronoun systems'. However, if pronouns, according to the very same source, "have little or no intrinsic meaning or reference" (*ibid.*, p. 221), what 'distinguishable entities' do they correlate with and what *is* the function of the category of number in the case of pronouns?

To an objective realist who believes that language represents aspects of the world which exists independently of the human cognizer, the world view shaped by his mother tongue has an objective character by default — therefore, it may be adequately reproduced in another language ("translated"). For example, if I discriminate between a thing and a number of things of the same kind, or between a substance and an instance of such substance by using special linguistic forms, it is just because language reflects, or 'represents' an aspect of objective reality via the grammatical category of number. Yet it is a trivial observation that what in one language is categorized as a thing may be categorized in a different language as a number of things or a substance. In Russian, such words as *gorox* 'peas', *kartoshka* 'potatoes', *luk* 'onions' are mass nouns, while their English counterparts are regular count nouns in the plural form. Names for distinguishable entities within a mass or substance such Russian nouns refer to are derived morphologically from corresponding mass nouns:

(12)

Mass Noun	Count Noun
gorox 'peas'	*goroshina* 'pea'
kartoshka 'potatoes'	*kartoshina* 'potato'
luk 'onions'	*lukovitsa* 'onion'
pesok 'sand'	*peschinka* 'a grain of sand'
sneg 'snow'	*snezhinka* 'snow-flake'

This is indicative of the different paths taken by conceptualization in the two languages: while in English peas, potatoes, onions are viewed as things that may be counted, in Russian they each are a mass in the first place, and a special word is needed to refer to a countable instance of such mass conceptualized as a thing. Substances such as sand or snow seem to be conceptualized similarly both in English and Russian, but while in Russian corresponding count nouns are derived from mass nouns and have the meaning 'a single distinguishable piece of the substance named by the root of the word', in English such meanings are lexicalized in a very different way — which, again, tells us that speakers of English and Russian seem to use language (allegedly governed by Universal Grammar) to describe different 'objective' worlds. Again, one and the same word may function as a name for a substance, that is, as a mass noun, and a name for a thing, becoming a count noun, as in *cheese — cheeses, wine — wines, water — waters*, even if in the latter case, as illustrated by example (13), it is hard to interpret the meaning of the plural form as 'a number of distinguishable instances of a substance':

(13) *The stately Amazon carries its wide waters to the Atlantic.*

It seems that the idea of plurality itself (just as many other concepts expressed in grammatical categories) is contingent on some specific mechanisms which are at work in the process of conceptualization of the so-called 'objective world'. Human cognizers speaking different languages 'partition' the world in different ways. So plurality as a concept, as a grammatical property characteristic of certain classes of words, does not reflect some universal 'truth' about the objective world, it reflects the fact that humans 'partition' it with the help of language, and they do so in the manner afforded by the language they speak — which often means that, as a Russian saying goes, "one can't jump over one's head". This is, by the way, one of the reasons why orthodox theory has been unable to understand and explain the phenomenon of language: it has been guided by common sense, or what Benjamin Lee Whorf (1956) called the 'natural language logic', overlooking the fact that humans exist in language which creates the world they live in (see also: Humboldt 1820; Maturana 2000; Imoto 2004b; Kravchenko 2007).

Taking a Gibsonian ecological approach to perception (Gibson 1979), we may speak of *affordances* provided by language. According to Edward Reed (1991: 140), human cognition is the "collective appropriation of affordances" taking place in an environment in which one can act to explore its action-perception possibilities. Bert Hodges (2007) extends Reed's theory of social affordances to utterances when he says that in talking with each other (and ourselves) we create affordances, opportunities that invite the other into seeing and moving in certain directions that look promising. When communicating with others, we create affordances in which linguistic aspects of our behavior are deeply integrated with the many cognitive processes characteristic of a living system, both second-order (an individual human) and third-order (a human community). As Reed (1996: 169) emphasized, "it is groups of humans, not individuals, that most typically embody the cognitive processes". In this, cognition is distributed, and so is *natural language* as central to human cognition.

In its attempts to understand how natural language works traditional linguistics has put a lot of effort into analyzing discourse as a connected series of utterances by one or more speakers. Yet, despite a long history of studies and a variety of related theories, the problem of discourse interpretation is still looming large and defiant on the linguistic horizon. As noted by Robert Trask (1993: 84),

> [O]n the whole, attempts at extending the methods of grammatical analysis to the study of discourse, in the hope of constructing 'discourse grammars', have not been outstandingly successful.

And this should not be surprising. Orthodox linguistics with its written language bias has erroneously assumed *utterance* and *sentence* to be synonymous, when sentence was defined as "any utterance or *written sequence* of words which is regarded as capable of standing alone to express a coherent thought" (Trask 1993: 250; emphasis added. — A.K.). The point, however, is that utterances do not express thoughts, they are signals which orient communicating parties in their consensual domain of interactions, they are cues for constructing meaning, and as such they *never stand alone* — they are always integrated with the multifarious aspects of the physical context in which they occur as well as with the inner states of the communicants as structure determined organisms each with its own history of ontogenetic structural coupling with the environment — and all these contribute to the construction of meaning. While natural spoken language is distributed over different time-scales, written texts, being graphic artifacts, are *not distributed over time,* and this is probably why persistent attempts on the part of linguists to describe oral discourse in terms of grammar as a set of rules used to combine written words into sentences into texts, have not been particularly fruitful.

But if this is something that cannot be done, are we justified at all in our use of the word *grammar* when we focus on regular patterns in the communicative behavior of humans? What is the purpose of grammar, anyway?

5.3. The Purpose of Grammar

5.3.1. The Sign System of Spoken Language vs. the Symbolic System of Written Language

Grammar as the art of writing is essentially different from the cognitive mechanisms governing discourse. Sentences are structural units of disembodied written texts, and as such they are symbolic artifacts. Being second-order cultural constructs, they are not grounded in first-order activity. As symbols, sentences are grounded in second-order linguistic activity such as speaking about sentences in an educational setting, for example, in an English class at school. A sentence is a concatenation of special graphic images (symbols) designed to represent, often in quite an arbitrary fashion, acoustic phenomena we call "words" which serve as orientational signals in our interactions with others. As Per Linell (in press) emphasizes, the individual constantly relates to others and to the environment. Because of their orientational impact, words are believed to possess meanings — at least, this is what common sense tells us. In reality, however, in spoken communication meanings are constructed on the spur of the moment by assigning orientational values to physical words which are integrated in the dynamically complex behavior of humans distributed over space in real time. According to Linell, sense-making (understanding) is neither contained in a subjective 'inner world', nor in an objective 'outer world', it lives in the relations between these, in an 'inter-world' between the organism and the external world. In other words, it is based in our experience of previous interactions in our cognitive domain including various kinds of activity such as first-order activity, first-order linguistic activity, and second-order linguistic activity. At the same time, it also involves online 'processing' of all the perceptual data provided by the physical context in which linguistic interactions take place. That is why in natural spoken communication there may be no such thing as 'purely linguistic data'.

The original purpose of spoken communication is to produce an immediate orientational effect on the protagonist. Such orientational effect may become manifested in one or several aspects of his behavior, both linguistic and non-linguistic. As argued by Linell (*ibidem*),

> [I]f languaging and thinking are other-oriented, it means that actions and cognitions are both responsive to prior events and actions, and anticipatory to possible next actions (on the part of oneself and the other). This in turn means that there is a close relationship between understanding and responding. The distinction between (utterance) production and comprehension is not as radical as has traditionally been assumed in models of senders (speakers, actors) and receivers (listeners). Preparing a response also involves stance taking and perspective-setting, and we can, at least in part, take the other's perspective.

Because of an intricate scaffolding provided by the physical context of a particular discourse occurring in real time, a behavioral response which we interpret as understanding may be obtained sometimes even *before* the speaker has completed a particular linguistic interaction aimed at orienting the other in his consensual domain. This is one of the reasons why utterances often differ in structure from sentences, despite the traditional belief that the latter are representations of the former to which they stand in one-to-one correspondence.

In new theoretical approaches to languaging as connotational behavior in a cognitive domain discussed in Part One, and counter to the common assumption that language is a set of fixed abstract forms, or, as Chomsky (1957) would have it, an infinite set of sentences, language is conceptualized in terms of potentialities, as something constituted by resources to be used in producing utterances and in associating understandings with utterances:

> These resources are designed to be completed only in situated meaning-making. They do not 'encode' or 'contain' their meanings; rather, they index, cue or prompt understandings in terms of reference, conceptualization and intervention (Linell, in press).

By contrast, in written texts understanding is not cued or indexed with reference to ever changing real-time situations in which communicating individuals find themselves, because texts, unlike spoken communication, are not intrinsically dialogical. As was pointed out in Part One, while languaging is orientational because, functionally, it is an extension of the human sensorium, written language is best viewed as a storage and retrieval system that allows humans to accumulate experience and knowledge. This is, indeed, where the concept of language as a sign system does come in, and this is where we may speak of its symbolic nature, having drawn a divide between the concept of sign and the concept of symbol. Elsewhere (Kravchenko 2003a) I have argued for a necessity to distinguish between the terms *sign* and *symbol* as essentially non-synonymous because the concepts behind them are different. The meaning of the Lat. *signum* is 'mark', while the meaning of *symbol* (from *symbolum* which is a later borrowing from Greek: συμβολον), is 'put together', 'thrown in'. Thus, one appears to be based in perception and experience, while the other presupposes an intentional stance. In Chapter 3 intentionality was defined as the capacity of a living system to modify its state of reciprocal causality with the world in order to sustain the ecological system which enables such reciprocal causality. With this in view, physical spoken words are *signs* — perceived empirical entities which function as 'marks' of other entities known from previous experience. As for corresponding graphic images we call words, they are not words understood as signs in the above sense, but *symbols* of such signs abstracted from all the specifics physical words as empirical entities possess in real space-time. Graphic images, such as letters of which written words are made, are just that which has been 'put together' and 'thrown in', and, once thrown into the relational domain of human linguistic behavior, written words begin to play a crucial role in human cognitive development, becoming, as the Russian writer Vladimir Soloukhin put it, 'a prosthesis for memory'. As symbols, written words sustain the ecological system formed by a third-order living system (a community of individual humans) and the environment, making up for possible breakdowns in the relational domain of natural (spoken) language.

5.3.2. The Orientational Function of Natural Language and Its Consequences for Grammar

In the course of natural language acquisition humans gain knowledge of the (semiotic) functions of linguistic structures experientially, through their behavior in a relational domain when they learn about the relations between one kind of empirical entities such as spoken words, phrases, and utterances, and all other possible kinds of entities and phenomena, including human behavior. It is in this relational domain that the orientational function of natural language emerges. As Charles Morris observed,

> [E]very sign involves behavior, for a sign must have an interpretant, and an interpretant is a disposition to response. But the behavior which is involved in the sign as sign itself occurs within the context of the more inclusive system of behavior of the interpreter of the sign, and frequently within the context of a behavioral system

> involving a number of organisms. There arises in this way the problem of the relation of signs to the individuals and societies in which signs appear and operate. <...> Not merely do signs have a certain signification at a given moment, but they have this signification only within the particular life history of their interpreters; and their appearance affects for good or ill the further life history of these interpreters (Morris 1971: 269).

Because signs have meanings 'only within the particular life history of their interpreters', these meanings may never be universal; interpreters with different life histories (or, in autopoietic terms, with different histories of fine structural coupling) will construct different meanings upon encountering similar signs as 'marks' of something which has been given in previous personal experience — at least, to the extent to which their individual experiences of interactions in the consensual domain differ. Because natural language is distributed over space and time, and because communities of individuals speaking the same language are also distributed across space, often over large territories stretching across different geographic and climatic zones, for any two individuals who speak the 'same' language but are, territorially, at the opposite poles of the bio-socio-cultural space occupied by the community, their consensual domain of interactions will consist in linguistic interactions with other members of the community dispersed over the entire territory between the poles. As a third-order living system, a community of individuals is sustained as a unity of interactions through the relational domain of linguistic behavior, which may be schematized as shown in Figure 5, where human individuals, each in his specific physical environment (solid circle), establish their consensual domains of interactions with others (overlapping circles), including linguistic interactions (overlapping dotted ovals) which create a relational domain which surpasses the physical boundaries of every individual's environmental niche. It is this relational domain of linguistic interactions in which, as Maturana points out, the process of recursive structural coupling takes place through the recursive application of the components of a consensual domain without the consensual domain. This is a very important idea in understanding languaging as a special dimension of the cognitive domain of interactions: an individual may interact with another individual in their consensual domain (the domain of interactions as interlocked conducts resulting from ontogenetic reciprocal structural coupling between structurally plastic organisms), or he may interact with another individual when there is no consensual domain, and do it with the help of language. Linguistic interactions are components of the consensual domain as they are grounded in the physical context of their occurrence; because of this, in the course of communicative interactions linguistic resources 'index, cue or prompt understandings in terms of reference, conceptualization and intervention'. Utterances do not function autonomously; as a sign, an utterance functions as an index which refers to its object because "it is in dynamical (including spatial) connection both with the individual object, on the one hand, and with the senses or memory of the person for whom it serves as a sign, on the other hand" (Peirce 1932: 170). By interacting with another individual outside his consensual domain, a speaker establishes what Maturana calls a second-order consensual domain, the relational domain of linguistic interactions in which the components of a consensual domain are applied without the consensual domain. It is a second-order consensual domain because utterances, being grounded in first-person experience, orient each of the communicating individuals with respect to their consensual domains. In doing so, they help establish common ground for understanding by referring to similar individual experiences.

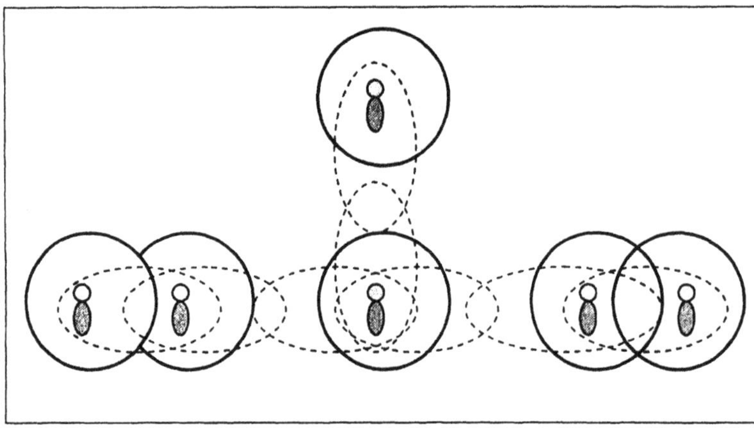

Figure 5. A community of individuals as a unity of linguistic interactions

As may be seen in Figure 5, the relational domain of linguistic behavior which sustains a community of individuals as a third-order living system depends on uninterrupted continuity in space-time; should a considerable number of links between individual humans operating in their domains of communicative interactions be severed for a prolonged period of time, the unity of interactions which characterizes a community of individuals as a (third-order) living system might cease to exist, marking the onset of disintegration of one living system into two or more smaller systems, as shown in Figure 6. Human history gives many examples of such processes when disintegration of the unity of the relational domain of linguistic interactions led to emergence of dialects which later developed into new languages with the consequent emergence of new communities/cultures/nations

— suffice it to mention the fate of the Latin language spoken in the Roman Empire and the emergence of modern Romance languages as its successors.

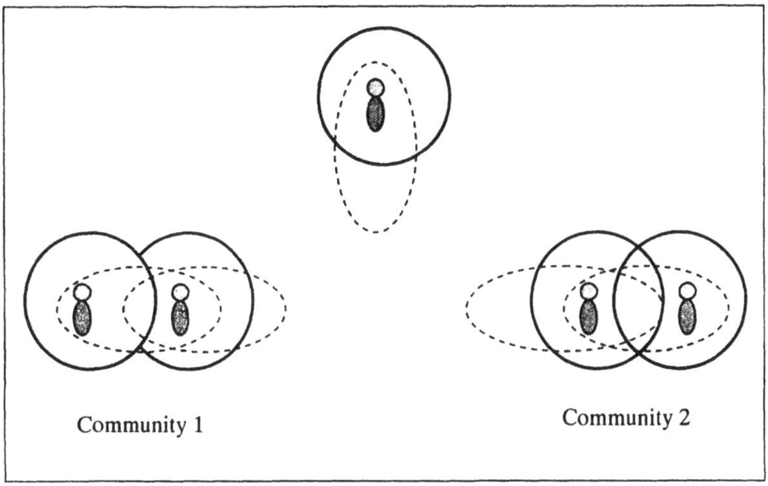

Figure 6. Disintegration of a third-order living system as a unity of interactions

What language enables us humans to do, and what in fact does make it an extension of our sensorium, is that by establishing a relational domain of linguistic interactions we become capable of monitoring and controlling aspects of physical environment which may be far beyond our actual physical reach, but which become accessible through the semiotic medium of language which orients us in our interactions with others while we, by return, manipulate others' behavior, and those others manipulate yet others' behavior so that in the end the environmental niche of someone at the opposite pole of the relational space may change because of our actual linguistic conduct; truly, as Morris observed, appearance of signs 'affects for good or ill the further life history of their interpreters'. But, of course, this is possible only if recursive communicative behavior we call 'languaging' is structured in such a way that the reciprocal causal relationships between concatenations of linguistic signs (such as utterances, for example) and other aspects of human behavior, being grounded in experiences common to different individuals because of a common history of interactions in a consensual domain, remain relatively stable across space and time. This stability is sustained by *patterned* linguistic behavior when linguistic recursion is causally related to other kinds of recursive behavior. And this stability of pattern gives rise to the concept of syntax as a description of the regularities in the concatenation of the elements of the consensual behavior. As emphasized by Maturana (see the quotation on p. 166), the syntax of any natural language is not a necessary set of universal underlying rules, it only refers to the universality of the process of recursive structural coupling. Just because many aspects of human cognition, particularly those related to sensorimotor activity and perception, are universal (we all have similar anatomies and physiologies, walk on two legs, etc.), they are bound to be reflected in language as an orientational signaling system, for what is orientation if not adjustment with relation to surroundings and circumstances, that is, positioning the self or other in space/time? And it is just because linguistic interactions are the recursive application of the components of a consensual domain without the consensual domain that, in order to function as orientational signals, linguistic signs must compensate for the lacking consensual domain by *indexing* aspects of the actual physical environment of the communicants. That is why natural linguistic signs *are indices, not symbols*, and that is why it is appropriate to speak of the phenomenology of signs and, by extension, of the phenomenological nature of spoken language (see Chapter 3). This epistemological necessity is succinctly summed up in the central thesis of the biological theory of cognition: "Everything said is said by an observer to another observer".

In a canonical face-to-face communicative situation linguistic signs prompt understandings of the communicative behavior of protagonists, while these understandings depend on many factors traditionally labeled as 'extralinguistic', such as the actual physical context in which communication takes place (including all interpretable perceptual phenomena), a history of interactions of each communicating party with its own environmental niche and a mutual history of interactions in a shared environment (that is, relevant experience), inner states (emotions, predispositions, etc.) and so on. In short, the meaning of a spoken utterance is not the result of interpretation of just linguistic signs alone or of their concatenations, it is constructed 'on-line' while integrating all the available 'extralinguistic' data which provide a multimodal real time scaffolding for understanding linguistic behavior. It is this feature of meaning-making which allows to speak of language as *distributed over space and time*. To put it differently, to view production and understanding of utterances in oral communication as linear processes determined by the linearity of concatenations of physical words — as traditional linguistics persists in doing to this day — is to make a categorical mistake bordering on blunder.

Because natural spoken language is distributed, its description in terms of grammar as a set of rules for organizing words into utterances is an ungratifying enterprise. Were it possible to formulate such rules, they would have to take into account all the characteristics of the ever changing situations in which languaging takes place — which, we must admit, is virtually impossible. Because a rule is always an abstraction of a kind, and because discourse interpretation always takes place in the concrete 'here-and now' of the communicating parties, there may be no universal rule which might take into account all possible parameters of all possible situations in which a particular linguistic sign might be used. For this reason, it makes no sense to speak of an emerging symbolic function of linguistic signs in the course of natural language acquisition (the so-called 'symbol grounding problem'), as at this stage there is no intentional 'throwing in' of new concepts in regard to the system of sign relations which already exist in the domain of linguistic interactions. Symbols do not have addressees; they are not communicative because a symbol cannot orient another individual. But what, then, about written texts?

If we compare the mechanisms underlying interpretation and understanding of written texts as compared to oral speech, we will see a number of crucial differences, some of which were mentioned in Chapter 1. While discourse is dialogical interaction in real time based on environmental affordances which consist in integrating different aspects of cognitive behavior (perceptual, sensorimotor, emotional, proprioceptive, linguistic) of the self and other, writing or reading a text does not imply or presuppose interaction with the other in *principle*. As was argued in Chapter 4, written texts are symbolic artifacts which consist of graphic images such as letters; these function as code signs for (clusters of) sounds that physical words are made of. It was also argued that isolated concatenations of graphic symbols we call written words do not represent physical words as a specific dimension of communicative behavior with the function to orient others (and oneself) in the cognitive domain of interactions. What they do is *symbolize* physical words as empirical entities we call natural linguistic signs. Signs of natural language are experiential phenomena and as such they are always grounded in first-person experience: a sign is always a sign to someone. That is why a written word does not symbolize something which is outside and above the concrete experience of a concrete language user (the so-called invariant), it serves as a scaffolding device, a 'prosthesis' that helps restore a relevant part of subjective experience from a person's memory. And if a particular concatenation of symbols, by analogy with other similarly looking concatenations of symbols which belong to the same code — such as the English alphabet, for example — is recognized by the reader as a written word, he may more or less correctly utter the sequence of sounds represented, or encoded, by these written signs, but these sounds would not have any particular meaning in the sense that they would not help restore from memory a fragment of relevant experience associated with the acoustic image just produced. The meaning of a word is its use, and if a language user has never heard others use a particular word in a particular situation (or situations), nor used it himself in the course of his personal history of fine structural coupling, to him it will simply have no meaning and produce no orientational effect. This is what, in fact, happened in the experiment described in Section 4 of Chapter 4.

If, however, while reading a text the reader encounters only symbolic representations of physical words each and all of which are part of his actual accumulated experience of interacting in his cognitive domain as a languaging observer, they *will* have a certain orientational effect on the reader, even though this effect will differ essentially from the orientational effect of similar physical words uttered in the course of spoken dialogical communication. Writing is asynchronous discourse, it is not a dialog with another human in real space-time. At best, as, for example, in the case of personal diaries or other texts of introspective character, writing may be viewed as a dialog with the self when its cognitive purport is not the same as when talking to others. Because we as human beings happen in language, speaking to oneself helps in defining our identity as conscious cognizers operating in the relational domain of linguistic interactions. Writing as a symbolic representation of our self-addressed speech, as a frozen replica of something that had a unique meaning at the moment of its occurrence, enables our recollection of that meaning, it allows us to re-evaluate it from a different point in time when we have changed as structure determined organisms, and depending on the result of such re-evaluation we may come to a new understanding — just as when we, by recalling someone's words uttered at a particular time in a particular situation, sometimes begin to 'see' them as meaning something different from what we initially perceived them to mean. Certainly, in such cases we are fully justified in speaking of the orientational effect of writing.

Again, as was shown in Chapter 4, the story is very different when ordinary language users view writing as representation of spoken language to which it is in one-to-one correspondence. As a result, communicative dysfunction arises when language users resort to writing in their attempts to orient others (intended or potential readers) as if they were in the here-and-now of the writer. Being a symbolic system, writing is not addressed to anyone, and the fact that we are so much used to viewing written texts as 'messages' to someone sent by their authors across space and time, does not change the truth of the statement made above: *symbols do not have addressees*. Nor can texts interact with one another, as is sometimes claimed, neither do they have a life of their own. They are what they are: lifeless artifacts, 'prostheses' for memory, a scaffolding that helps humans distribute the cognitive burden by unloading it onto something which, quite conveniently, may be always at hand. This cognitive burden is none other than categorized life-time experience: cognition is categorization (Harnad 2005b), and in the case of humans categorization is done in linguistic form. Taking space apart in the process of categorization, by 'partitioning' the world with the help of language, we attribute structure to it which facilitates our orientation in the world at hand, our *Umwelt*. Thus we come to the role of language in the generation and representation of knowledge.

5.3.3. Grammar as Knowledge Structure

As pointed out by Maturana (1970), the fundamental epistemological premise in approaching the problem of cognition consists in admitting certain facts as obvious truths about human cognition. Human cognitive ability is determined by his biological unity; besides, man is conscious of his knowledge. Being, fundamentally, a psychological and thus a biological function, consciousness specifies man's activities in the universum, and because of knowledge he is confident of his actions. 'Objective knowledge' seems possible, and thanks to 'objective knowledge' the universum begins to appear predictable as a system. However, knowledge as an experience is something personal, private which cannot be transferred to others. That which, as is believed, can be transferred, that is, 'objective knowledge', must be always construed by the hearer himself: the hearer understands, and the impression that 'objective knowledge' has been transferred to him appears only when he is prepared to understand. By emphasizing the role of the hearer in creating the so-called 'objective knowledge', the biological theory of cognition highlights the fact that the process of creating 'objective knowledge' is nothing but interpretation of linguistic signs which takes place in a spatio-temporal context informatively grounded against the hearer's knowledge, assumptions, and expectations. Therefore, the essence of cognition as a biological function is such that an answer to the question, "What is cognition?" must arise from understanding knowledge and the cognizing agent, and such understanding emerges from the latter's unique cognitive ability to partition the world with the help of language.

Ontologically, natural linguistic signs are a kind of category, a class of empirical phenomena singled out by the observer in his domain of interactions. These are further categorized into *kinds of signs* depending on the kind of semiotic relationship between a sign form (which may be simple or complex) and that to which the sign is perceived to be in a reciprocal causal relation. A particular kind of observers such as linguists engaged in second-order linguistic activity (professional linguistic discourse), go further and speak of the relationships between signs and what they stand for (the conventional domain of semantics), between different signs viewed as a system (the conventional domain of syntax), and between signs and their users (the conventional domain of pragmatics). So, in the case of linguistic categories one may speak of *categorization of categorization*, when an entity, already being a category of a kind, i.e., a significant entity, is used to express a category of another kind, becoming a signifying entity. Take, as an illustration, grammatical categories constituted by formal oppositions of signs of similar nature: every member standing in opposition to the other member expresses both general categorical and specific categorical meaning on condition that it is analyzed in the formal context of this kind of contrast. A change in the formal context which profiles a particular contrast may result in the change of meaning attributed to the form (which, typically, is described in terms of grammatical homonymy), just like a change of context may result either in emergence or disappearance of significance as a classificatory feature of an entity. It is for this reason that the immediate environment of linguistic sign in non-inflectional languages like English plays such an important role in sign interpretation.

However, if we assume that the meaning of a grammatical category is constituted by regular juxtaposition of sign forms with a particular type of context constituted by other signifying entities, then, as a consequence, the categorical meaning will be a feature not of a particular sign form but of an intersign relationship which, as a result, acquires semiotic status. It means that a grammatical category *is a kind of sign*. This thesis involves a certain taxonomic difficulty pointed out by Juri Stepanov:

> Since categories pose as something superordinate in regard to the forms of a given language, it follows that, as signs of signs, or metasigns, they must have names somehow different from the signs of this language. But because denotation is effected with the help of the very same language, categories are unavoidably named by the words of this language. This is a serious inconvenience that may result in misunderstanding (Stepanov 1981: 124ff.).

To linguists preoccupied with explaining language with the help of the very same language it certainly is a problem. But what about ordinary language users? In their blessed ignorance they are saved the trouble of looking for names for what they use as a means to achieve whatever they aim to achieve when they engage in linguistic communication. Just as when looking for pieces of dry wood to make a fire we are usually not concerned with the names of trees those pieces came from, neither are we concerned with the names of linguistic resources we use to produce a desired orientational effect on others. Guided by a lifetime of experience of recursive linguistic interactions in our cognitive domain, we choose particular sign forms in a particular situation knowing that we may quite reasonably anticipate a particular effect on the individual to be oriented — all because we have *knowledge* of how particular linguistic resources usually work in particular situations. Thus *knowledge is categorized experience,* and knowledge of grammar is categorized experience of intersign relationships which function as metasigns.

Once we define a grammatical category as a kind of sign, certain consequences follow which must be taken into account, the most important being that any grammatical category may be characterized as prototypically close to one of the three Peircean types of signs — icon, index, or symbol. Thus, certain types of syntactic structures are typical icons (Langacker 1988), the categories of verbal tense and aspect are typical indices (Kravchenko 1992), and the categories of nominal gender or case are typical symbols. At the same time, it should be borne in mind that, being semiotic phenomena, grammatical categories do not have rigid boundaries, they are fuzzy sets which often display features of 'family resemblance' — that is why linguists have always had problems with drawing a complete list of discrete grammatical categories found in a particular language.

Traditionally, grammatical categories are singled out and identified by analyzing written texts from the point of view of relations between signs, such as written words, phrases and sentences, and when a particular sign is found to be used recursively under certain conditions (which are described in terms of the context provided by other signs preceding or following the sign in question), linguists speak of a category and try to formulate a grammar rule that underlies the use of this category. However, the nature of grammar understood not as a set of rules for using signs but as some general principles underlying the mechanisms for reflecting interactional experience of the world in semiosis, cannot be explicated without addressing the issue of human cognitive activity, particularly, perception.

By establishing a second-order consensual domain (relational domain of linguistic interactions), communicating individuals orient one another in regard to their first-order consensual domains. This becomes possible due to perceptual groundedness of linguistic signs, including grammatical categories as metasigns; as empirical entities, linguistic signs are components of the environment as part of the organism-environment system, and grammatical categories as signs of signs reflect the relationships between signs which, in their turn, reflect the relationships between the cognizer and those aspects of the environment which, to the cognizer, stand in a reciprocal causal relation to signs of natural language. Experientially, spoken utterances, by partitioning the world of which they themselves are an essential part, *help construe and organize this world as a structured system of categorized experience.* In the case of natural language this structured system, being perceptually grounded in individual experience of multi-modal cognitive interactions with the world, integrates all aspects of such experience, both conscious and unconscious. But as soon as we turn to writing as a symbolic artifact, most of the cues and indices provided in spoken communication by 'non-linguistic' aspects of the consensual domain, without which interpretation and understanding become problematic, disappear. To make understanding of written language possible, the loss of the scaffolding in the form of a first-order consensual domain must be compensated for, and this is where we meet grammar — not as 'the system by which the words and morphemes of a language are organized into larger units', but as *a symbolized sign system for storing categorized cognitive experience (knowledge) wherein every grammatical category relates to some essential aspect of cognitive processing*. Accordingly, an analysis of linguistic categories both on the morphological and syntactic levels should aim at explaining the relationship between them and the cognitive processes which affect the resulting structure of a sentence (text) by imposing constraints on its organization.

Acquisition of natural language grammar is a natural autoreferential process in the course of which a child develops the ability to categorize perceptual data as concepts whose structures incorporate causal relationships between linguistic and non-linguistic objects, and to assign them specific cognitive values (see Chapter 3 for discussion). These values are the experiential outcome of interactions with the world, and they determine what is known as the speaker's linguistic competence because, ultimately, the starting point (point-of-reference) in the formation of any concept, however complex, associated with any linguistic structure regardless of its complexity, is *sensory data input*. The process of natural grammar acquisition relies heavily on the trial-and-error principle in building the experiential sign-object and sign-concept data bases which are, in fact, interconnected networks of representations as states of relative neuronal activity. These two data bases account for the binary principle in the organization of grammatical categories. Consequently, following John Goldsmith and Eric Woisetschlaeger (1982), two types of knowledge are distinguished as different cognitive values of grammatical categories: phenomenological and structural. Phenomenological knowledge is the result of identification, based on conceptualization and categorization, of observed objects and phenomena in the cognitive domain of interactions; it may be described as something that belongs to the epistemological level of the Observer. Structural knowledge is the result of categorization of linguistic interactions in the speaker's cognitive domain. Because languaging is the application of the components of a consensual domain without the consensual domain, because it is a *second-order* consensual domain distributed over space and time, an individual observer/speaker categorizes it as part of the generalized experience accumulated by past generations and "stored" in linguistic structures (recursive patterns) used by a community of speakers. Because structural knowledge belongs to the community as a whole, to any given individual speaking the language of this community it appears as something existentially given; so it may be described as something that belongs to the epistemological level of the Speaker. Since the speaker is always the observer, but the observer is not always the speaker, the distinction between the two types of knowledge is not always rigid and clear-cut, but it does have a systemic status and constitutes an important part of many grammatical categories often described in traditional grammar as a means for expressing evidentiality (cf. Willett 1987).

As a symbolized sign system for storing categorized experience, as a 'prosthesis' for memory, written language operates on two levels of knowledge representation: the level of text and the level of linguistic units used to produce a text. To the reader, the resulting in-formative value of a text — that is, its ability to have an orientational effect — stands in direct relationship to the in-formative value of physical words, phrases and utterances, as well as intersign relationships, grounded in the reader's experience and symbolized as graphic images which constitute the text. Therefore, to understand a written text the reader must relate his personal experience of individual signs of natural language symbolized by the discrete units of the text (phenomenological knowledge) to the structure of the text, or its grammar, which determines how phenomenological linguistic experience of the reader is construed by him into a possible world as a result of text interpretation. What it means is that the reader assigns experientially grounded values, acquired both in the course of first-order activity and first-order linguistic activity, to linguistic items symbolized in written form — such as words and constructions — and tries to make sense of these by organizing chunks of available personal experience according to the schemata provided by the way written words and constructions are organized into sentences. Since the reader's experience may very little, if at all, overlap with the writer's experience, the reader's understanding of a text as a result of his interpretative effort will never be identical to the writer's understanding of the same text — which, by the way, explains why literary criticism, as a way of earning a living, will never cease to be an attractive occupation — at least, to some languaging humans.

5.4. Summary

As I hope to have shown in this chapter, the concept of grammar as used in traditional linguistics needs to be re-evaluated and recast in a new mould provided by the epistemology of autopoiesis. Because of the written language bias in linguistics, not only is grammar far from being a clearly defined concept, but it is also misleading inasmuch as the purpose of grammar is understood to be the setting of procedures (rules) for organizing words and morphemes into larger units such as sentences. The logic behind such understanding of grammar leads to admitting that words and morphemes exist 'out there', that they have meanings, and that these meanings are combined and re-combined to express thoughts. In other words, the traditional conception of grammar is based on the same old myths linguistics lives by, and it is not surprising that little progress has been made in applying the theory of grammar thus conceived to solving practical tasks — at least, as they are seen in the traditional paradigm, such as writing a comprehensive guide to the grammar of a particular language with a complete inventory of grammatical categories and underlying rules, or the development of grammar checking utilities by the computer software industry, for example. By confusing natural language as a kind of communicative behavior, integrated in the complex dynamics of human cognition, with writing as symbolic representations of physical words which are orientational devices used in dialogical interactions in real space-time, orthodox linguistics makes a categorical mistake of viewing regularities in the use of written signs as rules which underlie the use of natural language.

Guided by common sense, orthodox linguists believe that the regularities observed in the linguistic behavior of others are 'independent' of the languaging subjects, just as mainstream cognitive scientists believe in the autonomy of cognition. The legacy of Cartesian dualism in thinking about natural language grammar leads to an assumption that grammars of particular languages are various realizations of the so-called 'universal grammar' as the necessary set of underlying rules common to all human natural languages. However, being structure determined organisms, humans display the capacity for recursive structural coupling which depends entirely on the operation of the nervous system as a closed neuronal network, that is why, as a set of coordinated procedures, the grammar of a human language may be any because its determination is contingent on the history of consensual coupling and is not a necessary result of any necessary physiology.

At the same time, although 'universal grammar' as conceived by generativists is a fiction, there is no denying the universality of the process of recursive structural coupling that takes place in humans. Because such structural coupling takes place in a consensual domain of interactions, and because human individuals belong to the same biological species and have similar physiologies regardless of the language they speak, phenomenological aspects of their interactions with the environment are necessarily characterized by similar patterns and co-variance, and they necessarily affect their linguistic behavior. Natural linguistic signs are not symbols, they are indices that cue human understanding in oral communication with regard to all the multifarious aspects of the physical context in which communication occurs, but particularly, with regard to the speaking observer as the ultimate point-of-reference. Such indexing is, basically, the cognitive function of many, if not all, grammatical categories in natural human language. These cuing aspects of natural linguistic signs underlie what linguists call *grammatical meaning*, and thanks to grammatical meaning expressed in grammatical categories as metasigns, a second-order consensual domain is established in which the components of a consensual domain of interactions are recursively applied without the consensual domain. The unity of a community of human individuals as a third-order living system is sustained by an uninterrupted continuity of the second-order consensual domain. In regard to spoken language, the purpose of grammar is to construe and organize the relational domain of interactions — a second-order consensual domain in which we exist as humans and the components of which we use to describe what we call the 'world' — by partitioning it with the help of linguistic signs grounded in first-order activity.

In writing, which is a system of symbols for representing the physical shape of spoken words as components of the consensual domain without the consensual domain, understanding may be cued only with the help of symbolized linguistic items themselves organized in such a manner as to compensate for the absent physical context by indexing points-of-reference used in simulating a second-order consensual domain. In regard to written language, therefore, grammar is a symbolized sign system for storing categorized cognitive experience (knowledge) wherein every grammatical category, or metasign, relates to some essential aspect of cognitive processing in a first-order consensual domain. How it is done is illustrated by autopoietically driven analysis of some categories in the chapters that follow.

CHAPTER 6

ASPECT: WHERE RUSSIAN AND ENGLISH MEET

6.1. Introductory Remarks

Verbal aspect as a grammatical category has received, alongside with tense, probably the most attention over the observable history of linguistic research both theoretical and applied. Yet, by and large, tense-aspect studies (at least, in Russian and English) have proved incapable of offering a relatively simple framework for treating these categories in a manner *comprehensible to a linguistically unsophisticated person* who, lacking formal training in linguistics, wishes to learn a foreign language. Although "in an explanation, simplicity is power" (Clark and Chalmers 1998), applied grammatical frameworks notoriously lack lucidity in addressing instructional tasks in second or foreign language acquisition.

This inadequacy is rooted in analytic philosophy exploited by orthodox linguistics as its methodological foundation, with its propensity for logically biased semantic analysis as the ultimate goal of linguistic study. The role played by logicians in setting up formal frameworks for the analysis of meaning in natural language, especially in the first half of the 20th century, can be hardly underestimated, and their legacy is still palpable in accounts of those parts of grammar which are believed to be more or less directly related to the aspects of 'objective' phenomena in the 'real world', such as time, for example. Characteristically, aspect has been systematically treated as a grammatical category whose meaning is related to internal temporal structure of a situation. Thus, the grammatical meaning of aspect has been implicitly viewed as secondary in relation to tense because situations possessing temporal structure don't exist outside discourse organized along the so-called time axis which schematizes human notions about 'objective' time. The internal structure of a situation, categorized in linguistic form, has been described in terms of perfective, imperfective, habitual, durative, punctual, iterative, etc. aspects. Nevertheless, all such fine-grained classifications fail to do the job of helping a learner of a foreign language understand the 'works' of grammar when it comes to tense-aspect acquisition, and one of the reasons why they do is that they can hardly be characterized as simple; therefore, they lack the explaining power.

Verbal aspect as a grammatical category is singled out in different languages due to a system of grammatical (morphological) forms whose meaning is not limited to reflecting the usual tripartite system of temporal distinctions such as PRESENT, PAST, and FUTURE, but also includes (at least, such is the belief) an additional characteristic feature of the process (activity, state) expressed by the verb, that is, the manner in which the process occurs or activity is carried out. The manner or character of the action flow and its distribution in time is considered to be a semantic categorical feature known as "aspectuality". However, the semantics of aspect continues to cause much dispute.

Aspect studies in different languages take as a starting point the classical aspectual system in Slavic, in which the basic distinction between what is known as PERFECTIVE (PF) *and* IMPERFECTIVE (IMP) *is morphologically sustained*. It is exactly at this point that the whole controversy starts, for there is a profound lack of agreement on what the actual meanings of these aspect forms are. One of the most acclaimed interpretations of aspect meaning is that based on the notion of *boundedness*: "Boundedness is the idea of completeness (exhaustion) of the temporal manifestation of the action as expressed by the verb" (Bondarko and Bulanin 1967: 47). Thus, the meaning of the Russian PF is defined as the totality of an action expressed by the verb, whereby the action is viewed as a spot-like, non-continuous event that reaches its bounds and whereupon a certain result of the action is obtained. The meaning of the IMP aspect is usually associated with continuity and linearity of an action in its occurrence, without any reference to the action's bounds *per se*, and with its processual and generic-factual function (Shvedova and Lopatin 1989; Xrakovskij 1990). This semantic approach was adopted by Indo-European linguistics and is reflected in the classification of verbs according to different aspectual classes (Vendler 1967), or the so-called Aktionsarten. Yet, as pointed out by Igor Miloslavskij, it is very likely that the whole controversy about the nature and meaning of aspect has been based on a fallacy:

> The paradox of the current situation in linguistics is that aspect as a specific set of features of a verb lexeme is singled out strictly and consistently on exclusively grammatical grounds (i. e., combinability and paradigmatic relationships. - A.K.). However, all the efforts of scholars have been directed at presenting aspect in such a way as if it were a category defined on strictly semantic grounds (Miloslavskij 1989: 39).

As a result, traditional semantic theories of aspect in Russian are far from being simple and/or comprehensible, so it is no surprise that for a foreign learner of Russian acquisition of aspect is a challenge that cannot be met in an instructional classroom setting.

In what follows I will not be concerned with the numerous current theories of aspect and their criticisms, since to do so would mean to continue the very tradition I want to break away from. Nor will I start by placing the analysis of aspect presented in this chapter within the theoretical and conceptual framework of the biological theory of cognition and language outlined in Part One. By doing so, I hope to refute possible accusations of being aprioristic. Rather, I will offer an unbiased approach to the phenomenon of aspect: taking a totally different angle, I will look, in the first place, at what in traditional linguistics is referred to as "purely linguistic data" — that is, grammatical forms of the verb as they are used in (written) texts. But first, I will consider the notions of boundedness and totality which, according to current Russian grammars, constitute the semantic invariant of aspectuality. I will show the vacuousness of these notions as something based on the faulty logic of rationalism, and dismiss them as absolutely untenable. Then, I shall look at some intriguing data from the Russian syntax and morphology and analyze the derivational patterns for base verbs (it is important to remember that Russian aspect is a lexical-grammatical category based in both lexical and grammatical derivation). These observations will be followed by a discussion where I hope to show that the true nature of aspect in Russian is radically different from what it has been traditionally believed to be. After that, I will turn to English and apply a similar procedure to the analysis of the so-called tenses. A contrastive analysis of aspect in such morphologically different languages as Russian and English will provide a good illustration of the cognitive universality of aspect as a means of grammatical categorization of *perceptually different types of situations*. It will be shown that in both languages the grammar of aspect is rooted in similar conceptualizations of events which are categorized relative to the *identifiable observer as a source of information*. Having done this, I will conclude by laying a bridge from the undeniable facts about the meaning and function of aspect in Russian and in English to the epistemological premise of the biological theory of language and cognition: "Everything said is said by an observer to another observer".

6.2. Aspect in Russian

6.2.1. Boundedness and Totality

Aspect studies in Russian have a long history, but until recently the semantic cornerstone of the theory of aspect — the concept of boundedness — has not been seriously questioned (Kravchenko 1995; 2004), although the plausibility of the assumption that actions expressed by verbs have any bounds imposed on them by the verb meaning was put to test by Juri Stepanov (1976). To begin with, the term 'action' used to refer to whatever the verb stands for, is not a very good or convenient one: it is not an action that is described by a verb lexeme, but rather an activity, or process of which an action is, or can be, but a part. The notion of 'action' belongs to syntax; an action is not the referent of the verb itself, it is the referent of the verb phrase. Outside the syntactic structure, the meaning of the verb is that of activity or process (or change, in a general sense) and nothing more; it does not imply a starting point or an end point, for that matter, so it cannot and does not have "inner bounds". Consider (14):

(14) *Ty prishla vovremja. Ostalos' dve minuty.*

You come PF PT in time. Remain PF PT IMPERS two minutes.

'You've come in time. We've got two minutes left.'

Traditionally, *prishla* is defined as the past tense of the PF verb *prijti* 'come' which means, in the parlance of semantically driven explanations of grammatical phenomena, that the action expressed by this form has reached its inner bounds and has exhausted itself in the sense of its total completion. This is the semantics of the perfective as it is treated in orthodox grammar. But let us imagine a situation in which an utterance such as (14) could occur. For example, I have to meet someone at a certain place at a certain time (let us say, it's a first night at a movie theater). My date, as is often the case, turns up at the very last moment. Even as she is approaching me, I start moving toward the entrance door, uttering (14). My date does not have a chance or time to stop; in fact, she may even be having to speed up a bit to keep up with me. The activity of going, which is the lexical meaning of the verb *prijti*, is not interrupted even for a moment, it is still taking place after I utter (14). On what grounds, then, are claims made that this very activity has exhausted itself reaching its inner bounds? Let us deliberate this question.

We may begin with a trite observation that an activity by itself, categorized in linguistic form as a certain type of 'real world' phenomena, does not have any inner bounds or limits. When we look out the window and see a moving object such as a running dog, for example, we know that the object is moving due to successive changes of its positions in space registered by our sensorium, but nothing in the process itself indicates that it started at a moment prior to the time of observation, or that it is heading toward an end at some moment after. We only see what we see — nothing more and nothing less. Any speculation about there being a starting point or an end point of the dog's run is based not on what we know about a particular kind of locomotion, but on what we know about dogs as an animal species which belong to a wider class of animate objects which constitute a part of our worldview. And in this worldview, dogs are simply living creatures which can be in a variety of states with regard to their motion in space: we can observe them as relatively motionless when resting or in sleep, or as moving in space in all kinds of manner (running, trotting, scampering, dashing, etc.). It is this knowledge (knowledge that the dog's *modus vivendi* is not continuous running) that enables us to reasonably assume that if we see a running dog, there must have been a preceding period or moment of time when the dog was not running. *Ergo*, there was a starting point to the dog's run, as well as there will be an endpoint, too. But what about rivers running to the sea? Can we speak of a starting point of a river's run? Our common sense tells us that if we cannot say about what we refer to as river that it *runs* or *is running*, then in all likelihood we cannot call it a river.

Crucially, it is not the verb itself that accounts for the meaning of completion of the activity referred to by the predicate, but our knowledge of how things, referred to by respective complements, come into being, as in the following examples:

(15) *Ivan narisoval krug.*

Ivan draw PF PT circle A

'Ivan drew (has/had drawn) a circle.'

(16) *Masha napisala pis'mo.*

Masha write PF PT letter A

'Masha wrote (has/had written) a letter.'

(17) *Miss Marple svjazala paru perchatok.*

Miss Marple knit PF PT pair A gloves G

'Miss Marple (has/had) knit a pair of gloves.'

In (15), starting with a curve and continuing one of its ends until it meets the other, Ivan does not necessarily have to interrupt the process thereupon. Suppose the lead of the pencil, or the tip of the brush, or whatever it is that Ivan is using for drawing, is very thin, and the picture he gets is rather light — it would be quite natural for Ivan — and many other people, for that matter — to enhance it by repeating the process as many times as necessary. Would an utterance such as (15), produced by an observer after the ends of the curve meet, mean that the activity of drawing was completed? Although the observer, uttering (15), might mean just that, could it be regarded as some indisputable 'objective truth' about the situation thus described? Obviously not; it would mean that in the process of the activity of drawing a certain recognizable object emerged, but surely this would not imply that at the moment this object came into being the activity, which caused its appearance, ceased. For all we know, the circle may be only a part of a more complicated graphic pattern consisting of other elements which Ivan can continue to draw without any pause; he may not even be thinking about it as a circle, and should someone ask him, "What did you just draw?" he might answer, "Nothing. I just got started."

In (16) the situation described is not necessarily such that the activity of writing is no longer present therein; Masha may still be writing something — another letter or suchlike — at the moment (16) is uttered. Are there sufficient grounds to assert that the activity of writing exhausts itself after reaching its "inner bounds"?

The process of writing a letter may be only a part of the activity referred to by the verb, a certain stage singled out on the basis of our knowledge of what it is that we refer to as 'letter'. In exactly the same manner, the situation described in (17) does not necessarily preclude a possibility for the activity of knitting to be still in progress when (17) is uttered. Miss Marple can be starting work on a new garment right after she has finished the second glove in a pair, and we can utter (17) even as we are watching her skillfully setting out a new base pattern for something she wants to begin to work on. In this case, the activity referred to by the verb ('knitting') is still present in the situation as observed by the speaker, so an assertion about the completion of this activity as resulting from reaching its "inner bounds" would appear simply false.

However, the above considerations do not apply to all perfective verbs. Other than prefixed, there are two other types of perfective verbs: non-derived, or base PF verbs and derived suffixed PF verbs. The number of base PF verbs such as *brosit'* 'throw', *past'* 'fall', *dat'* 'give', etc. is relatively small, and they are survivors from an earlier stage in the historical development of the Russian language. These are mainly verbs of motion and transfer in space; they refer to what may be called *phenomena of primary cognition* — as far as linguistic categorization of processes and events is concerned. In other words, concepts expressed by such verbs belong to the class of semantic prototypes (Jackendoff 1983; J. Taylor 1989; Geeraerts 1989 *inter alia*). Derived suffixed PF verbs (the so-called *-nu*-verbs) such as *prygnut'* 'jump' (from IMP *prygat'*), *dunut'* 'blow' (from IMP *dut'*), *metnut'* 'cast' (from IMP *metat'*) and many others, together with base PF verbs, make up the stock of purely grammatical forms that express the PF aspect — as opposed to lexical-grammatical forms in the case of prefixed verbs. These verbs, typically, name momentary events which are no longer observable at the moment the verb is used in an appropriate utterance. Consider examples (18) - (21):

(18) *Lisa brosila kuklu.*

Lisa throw PF PT doll A

'Lisa (has/had) dropped the doll.'

(19) *Lena dala mne knigu.*

Lena give PF PT me book A

'Lena gave (has/had given) me a book.'

(20) *Ivan prygnul vniz.*

Ivan jump PF PT down

'Ivan (has/had) jumped down.'

(21) *Ona metnula na menja serdityj vzgljad.*

She cast PF PT on me angry glance A

'She cast an angry glance at me.'

Brosila in (18) does not refer to the motion of the object, but to the manner in which this motion takes place, or to the action that causes this motion, this action being completed at the time of utterance. Similarly, the verbs *prygnul* in (20) and *metnula* in (21) describe the manner in which the respective actions are carried out, whereby the consequent processes — such as Ivan's motion from one place to another in (20), or the movement of her [head and] eyes in (21) — come into being. In (19), a change of ownership relation between the object and its possessor is described which, in a prototypical situation, is effected by the object losing physical contact with its former possessor and coming into physical contact with the new possessor. The form *dala* indicates that the above sequence obtains (i. e., the book is now in my hands), but it does not refer specifically to either one of the constitutive actions (yielding and taking), they are viewed as a homogeneous event preceding the utterance of (19).

As examples (18) – (21) show, the meanings of some PF verbs do seem to incorporate the semantic features "boundedness" and "totality", although these verbs constitute only a small part of the stock of PF verbs. The PF *-nu-* verbs seem to be a way of compensating for the loss of the majority of base PF verbs in Old Russian, so it is the base PF verbs that are of primary interest, as they appear to be authentic bearers of the semantic features "boundedness" and "totality". However, it remains unclear how these features in the meaning of base PF verbs can be accounted for. Nor is it easy to tell what cognitive factors play a role in the process of meaning-making which results in linguistic categorization of these features in a special grammatical form such as a base PF verb stem. And why are these semantic features not obligatorily characteristic of prefixed PF verbs? Looking for an answer to this intriguing question, let us now turn to the morphology of Russian aspect.

6.2.2. The Morphology of Aspect

In examples (14) – (17) we find PF verbs which are derived from corresponding IMP verbs by adding a spatial prefix. In terms of productivity, prefixation is the primary means of perfectivization of IMP verbs in Russian, and the number of prefixes that serve this purpose is quite large. The majority of such prefixes are locative in nature, and they often preserve their original lexical meaning, especially in verbs of motion, cf.:

(22) Spatial Prefix + IMP *idti* 'go' → PF verb

vy- 'out'	***vyjti*** 'go/come out'
u- 'away'	***ujti*** 'go away, leave'
pri- 'to'	***prijti*** 'come'
pere- 'over, across'	***perejti*** 'go across, cross'
pro- 'through'	***projti*** 'go through'
na- 'onto'	***najti*** 'find'
ot(o)- 'off, aside'	***otojti*** 'move/step aside'
s(o)- 'from'	***sojti*** 'step down from'
za- 'behind'	***zajti*** 'go behind', etc.

In examples (18) – (19) base PF verbs are used which, together with derived (suffixed) PF verbs such as in (20) – (21), illustrate what constitutes the purely grammatical phenomenon of the perfective aspect proper. Such verbs can take prefixes which modify their lexical meaning in terms of direction of the motion, but the grammatical meaning of the verb remains unchanged. Suffixed PF verbs, as well as prefixed PF verbs of the type illustrated in (20) – (22), are derived from imperfective verbs such as *prygat'* 'jump', *metat'* 'cast', *idti* 'go', etc. Such IMP verbs are non-derived, therefore, they must also be considered as basic. Thus, there are two types of base verbs, PF and IMP. Base PF verbs always have base IMP counterparts, but the opposite is not true as base IMP verbs yield PF verbs (the *-nu*-verbs) derived from IMP stems:

(23)	a.	**PF:**	*brosit'*	*dat'*	*past'*	*pustit'*
		↓	THROW	GIVE	FALL	LET
		IMP:	*brosat'*	*davat'*	*padat'*	*puskat'*
	b.	**IMP:**	*prygat'*	*dut'*	*idti*	*delat'*
		↓	JUMP	BLOW	GO	DO
		PF$_{der}$:	*prygnut'*	*dunut'*	*prijti*	*sdelat'*

This gives us an idea of the hierarchy between base PF verbs and their counterpart IMP verbs, and characterizes the PF verb as the strong ('marked') member of the category. Base imperfective verbs behave differently; there is a series of verbs in

Russian which form lexical pairs the same way as the PF/IMP pairs do, the difference being that both verbs in such a pair are non-derived imperfective verbs. Let us mark them for convenience as IMP_1 and IMP_2. These are mainly verbs of locomotion, such as in (24):

(24)

IMP_1	IMP_2	Lexical Meaning
idti	*xodit'*	GO
bezhat'	*begat'*	RUN
polzti	*polzat'*	CRAWL
letet'	*letat'*	FLY
vesti	*vodit'*	LEAD
katit'	*katat'*	ROLL
etc.		

The question prompted by the data is, "Since the lexical meaning ('specific kind of locomotion') of such paired verbs is the same, what are the two different forms for?" Several explanations have been suggested in the literature. Alexander Potebnja (1977) defined the meaning of IMP_1 as "concrete" and contrasted it with the meaning "habitual" or "regular" for the IMP_2 verbs. In Roman Jakobson's (1971) terms, the distinction is between the "determinate" and "indeterminate" aspects. In more recent literature these verbs have been contrasted either on the feature "directedness (IMP_1)/ non-directedness (IMP_2)", or on the double feature "duration/frequency". However, none of these approaches may be accepted as satisfactory. Consider the following examples:

(25) a. *Lena idjot v shkolu.*

Lena go IMP_1 PR in school A

'Lena is going to school.'

b. *Lena xodit v shkolu.*

Lena go IMP_2 PR in school A

'Lena goes to school.'

(26) a. *Ja idu.*

I go IMP_1 PR

'I'm walking.'

b. *Ja xozhu.*

I walk IMP$_2$ PR

'I walk.'

In (25), directedness is an obvious part of the meaning of the verb in both (a) and (b), while it cannot be said about (26a) and (26b). The "duration/frequency" contrast does not constitute an opposition in the strict sense of the term as two different features are contrasted. We could speak of a categorical opposition in the case of one of the following semantic contrasts: "durative/non-durative", or "frequentative/non-frequentative" — but then IMP$_1$ would be the marked member in the first instance and the unmarked member in the second, which doesn't really make much sense. Besides, in sentences such as (27a) and (27b) the idea of duration is inherent in the meaning of the verb:

(27) a. *On bezhit po dorozhke uzhe tselyj chas.*

He run IMP$_1$ PR along track D already whole hour

'He's been running on the track for a full hour already.'

b. *On begajet po dorozhke uzhe tselyj chas.*

He run IMP$_2$ PR along track D already whole hour

'He's been running on the track for a full hour already.'

Explanations offered by Potebnya and Jakobson are, apparently, similar, although it is not quite clear what is understood by "concreteness" or "determinate aspect". However, all these approaches are, at least partially, justified as based on a keen intuition of the language users who proposed them. Such features as "concreteness", "determinacy", "direction", "uration", and "frequency" bear, in a certain way, on the main categorical feature that underlies aspectual oppositions and which will be discussed later. However, besides the derivational patterns mentioned above, there is one more suffixation pattern that affects only base IMP$_2$ verbs capable of taking the *-yva-* suffix:

(28) *xodit'* → ***xazhyvat'*** 'go'

letat' → ***ljotyvat'*** 'fly'

delat' → ***delyvat'*** 'do'

etc.

The role of this suffix is purely grammatical, it does not affect the lexical meaning of the verb, nor does it change its aspectual status. Part of the meaning of the verbs with the *-yva* - suffix is reference to an event or process as something which takes place at a non-specified time in the past and is relevant for the current experience of the agent, that is, it is close in meaning to the experiential perfect in English. In Modern Russian, the *-yva-* verbs are used rather seldom and only in the past tense, gradually becoming obsolete (further marked up with a cross: H).

So, to sum up the 'purely linguistic' data, there are three groups of morphologically opposed verbs in Russian as shown in (29), and the distinction within each of them is not lexical but grammatical:

(29)	A.	IMP_1	IMP_2	IMP_3	MEANING
		idti	*xodit'*	H*xazhyvat'*	GO
		vesti	*vodit'*	H *vazhyvat'*	LEAD
		letet'	*letat'*	H *ljotyvat'*	FLY
	B.	PF_1	IMP_2	IMP_3	
		brosit'	*brosat'*	H *brasyvat'*	THROW
		pustit'	*puskat'*	---------	LET
		past'	*padat'*	---------	FALL
	C.	PF_2	IMP_2	IMP_3	
		prygnut'	*prygat'*	H *prygivat'*	JUMP
		tolknut'	*tolkat'*	H *talkivat'*	PUSH
		metnut'	*metat'*	---------	CAST

We may leave the few IMP_3 verbs registered in dictionaries aside in our consideration of the morphologically sustained aspectual distinctions as they are not part of regular oppositions. If we now compare the contrasted pairs in each of the three groups, it will become obvious that IMP_2 verbs are the weak ('unmarked') members of respective oppositions. However, if for Groups B and C their counterparts are marked for the so-called perfective aspect (allegedly expressing "boundedness" and "totality"), it is not so for Group A. What, then, is in common (if anything at all) between IMP_1 verbs on the one hand, and PF_1 and PF_2 verbs on the other? Depending on how this question is answered, the entire conceptual framework for the traditional theory of aspect may have to be revised (which, in fact, I intend to do).

All the aspect forms in (29) — IMP$_1$, IMP$_2$, IMP$_3$, PF$_1$, PF$_2$ — are base lexemes subject to further derivation by means of prefixation. Here we observe an intriguing phenomenon: what would seem to be a trivial case of lexical derivation suddenly results in a change of the grammatical status of IMP$_1$ verbs, such as *idti* 'go' (cf. example (22)), whereas in the case of IMP$_2$ verbs an added prefix only modifies the lexical meaning, but does not affect the grammatical meaning:

(30) Spatial Prefix + IMP$_2$ *xodit'* 'go' → IMP$_{2der}$ verb

vy- 'out'	**_vy_**_xodit'_ 'go/come out'
u- 'away'	**_u_**_xodit'_ 'go away, leave'
pri- 'to'	**_pri_**_xodit'_ 'come'
pere- 'over, across'	**_pere_**_xodit'_ 'go across, cross'
pro- 'through'	**_pro_**_xodit'_ 'go through'
na- 'on(to)'	**_na_**_xodit'_ 'find'
ot(o)- 'off, aside'	**_ot_**_xodit'_ 'move/step aside'
s(o)- 'from'	**_s_**_xodit'_ 'step down (from)'
za- 'behind'	**_za_**_xodit'_ 'go behind'
etc.	

Prefixed IMP$_3$ verbs in Group A also retain their aspectual (in the traditional sense) status, but the prefix sometimes loses its spatial meaning, as in **_vy_**_xazhivat'_ 'swagger', or the entire meaning of a prefixed verb may change due to metaphoric transfer, as in **_u_**_xazhivat'_ 'nurse'. In Group B prefixation is a purely lexical derivational procedure accompanied by a shift of accent; the base PF$_1$ verb retains its aspectual meaning whereas its lexical meaning is modified just like in the case of prefixed IMP$_2$ verbs:

(31)

Spatial Prefix	+ PF$_1$	→ PF$_{2der}$
vy - 'out'	*brosit'* 'throw'	**_vy_**_brosit'_ 'throw out'
u- 'away'	*past'* 'fall'	**_u_**_past'_ 'fall down'
za- 'behind'	*sest'* 'sit'	**_za_**_sest'_ 'sit in hiding'

IMP_2 verbs in Group B behave similarly, that is, with an addition of a prefix they do not change their aspectual meaning — although in some cases there are certain constraints, when a particular verb cannot take a particular prefix:

(32) IMP_2 + Spatial Prefix → IMP_{2der}

puskat' 'let'	*vypuskat', otpuskat', perepuskat'*
padat' 'fall'	*vypadat', otpadat', perepadat'*
davat' 'give'	*vydavat', otdavat', peredavat'*, etc.

Some verbs, for example, *brosat'* 'throw', seem to be an exception, as an added spatial prefix changes their grammatical status:

(33) IMP_2 + Spatial Prefix → PF

brosat' 'throw'	*zabrosat'*, *sbrosat'*, ***pere**brosat'*, etc.

The vacant slot of the transformed IMP_2 verb (and of the inadmissible forms such as **vybrosat'* 'throw out') in the aspectual paradigm is filled by the prefixed verb with the *-yva-* suffix, thus compensating for the loss:

(34) IMP_3 + Spatial Prefix → IMP_{3der}

* *brasyvat'* 'throw'	*vybrasyvat'*, *sbrasyvat'*, ***pere**brasyvat'*, etc.

Finally, in Group C the PF_2 and IMP_2 verbs are affected by prefixation in a way similar to the PF_1 and IMP_2 verbs in Group B, except when prefixation is precluded by combinatorial constraints. Our short survey of the morphology of aspect may be summed up as follows:

1. Aspect in Russian is a grammatical category sustained by regular morphological oppositions of two different verb forms which have the same lexical but different grammatical meaning.

2. There are three basic types of aspectual oppositions singled out on strictly grammatical (morphological) grounds:

(i) base PF_1 verbs vs. base IMP_2 verbs,
(ii) derived suffixed PF_2 verbs vs. base IMP_2 verbs,
(iii) base IMP_1 verbs vs. base IMP_2 verbs.

3. “Boundedness” and “totality” as semantic features associated with the grammatical meaning of the perfective aspect may be ascribed only to PF_1 and PF_2 verbs, but they cannot explain the meaning of prefixed PF_3 verbs which constitute the overwhelming majority of perfective verbs in Modern Russian.

4. IMP_2 are the weak, and PF_1 and PF_2 are the strong members of the opposition. It means that IMP_1 must also be marked for the feature shared by PF_1 and PF_2, the feature on which aspectual oppositions are based.

With this in mind, let us now look at data of two kinds: (i) grammatical (morphological) paradigms of the Russian verb, and (ii) functional constraints on the use of different aspects in discourse, and the nature of such constraints.

6.2.3. The Grammatical Paradigm of the Russian Verb

The Russian verb has a large number of grammatical forms: the finite tense-aspect forms, the infinitive, two participles (active and passive), each of which has two forms traditionally labeled as “present” and “past”, and two verbal adverbs. In this section, we will look at the grammatical paradigms of the three base-type oppositions as shown in (29) and at the effects of prefixation.

Table 4 gives an overview of the morphological forms of paired IMP_1/IMP_2 verbs, alongside with the obsolete form IMP_3 (the suffixed *-yva-* verb). As may be seen from the table, an almost complete paradigm is found in the case of IMP_2, with only one possible form missing, the past verbal adverb * *vodiv*. The IMP_1 paradigm is one member shorter, lacking the past passive participle with a resultative meaning, though it may be easily created in concordance with the derivational pattern. All IMP_1 verbs are subject to this constraint, while IMP_2 verbs yield present passive participles, for instance, *katat'* ‘roll’ → *katan, taskat'* ‘drag around’ → *taskan.* Even intransitive IMP_2 verbs such as *xodit'* ‘go/walk’, *jezdit'* ‘ride’ used to be passivized: Modern Russian still features adjectivized past participles *nexozhenyj* ‘not walked on’, *nejezzhenyj* ‘not ridden on’, and impersonal predicatives such as *xozheno,* as in (35) – (36):

(35) *Mnogo bylo xozheno po etoj tropinke.*

A lot be IMP_2 PT IMPERS walk IMP_2 PRED along this path D

‘This path had been walked a lot.’

(36) *Xazhival i ja kogda-to v restorany.*

Walk IMP_3 PT S and I sometime into restaurants A

‘There was a time when I, too, used to go to restaurants.’

IMP_2 past passive participles do not possess a resultative meaning; rather, their meaning is related to the concept "factual" based on empirical knowledge. IMP_3 verbs in Modern Russian have only one form — that of the past tense, and even then, not all of them, as is seen in Table 4.

The right side of the table shows what happens when a spatial prefix is added to the base verb. Only the IMP_2 paradigm remains unchanged: although the lexical meaning of the verb is modified in regard to the motion's direction, the derived verb still belongs to the same grammatical class IMPERFECTIVE. The IMP_1 verb becomes perfective and represents another morphological type, PF_3 (beside the base PF_1 and suffixed PF_2 verbs). PF_3 verbs consistently preclude any present tense forms. Moreover, the present tense form *vedjot* 'lead', combined with a prefix which retains its full lexical meaning, becomes a future tense form (*vyvedet*), and the past verbal adverb **vev*, missing in morphological paradigm of the IMP_1 verb, is restored in the prefixed PF_3 form *vyvev*. As for the IMP_3 verb, it fully restores its grammatical paradigm through prefixation, but its meaning ('play a hooked fish') becomes metaphorized and the verb thus falls out of the aspectual opposition proper.

Let us now look at Table 5. The morphological paradigms of the base IMP_2 and derived IMP_{2der} verbs are identical to the corresponding paradigms in Table 4, and the PF_3 paradigm of the prefixed verb in Table 1 is identical to the PF_3 paradigm in Table 5 which, in its turn, mirrors the base PF_1 paradigm. The verb *dat'* 'give' (as well as *past'* 'fall', *pustit'* 'let') does not have an IMP_3 form, while the verb *brosit'* 'throw' does. The form *brasyvat'* possesses an incomplete paradigm similar to that of the verb †*vazhivat'* 'lead' in Table 4 which, however, is fully restored as a result of prefixation. At the same time, the meaning of the prefixed verb in this case is not metaphorized, it is substituted by a meaning typical of IMP_{2der}, and the verb *vybrasyvat'* 'throw out' makes up for the missing verb * *vybrosat'*.

In Table 6, the picture is somewhat different. While the suffixed PF_2 and derived prefixed PF_{2der} paradigms correspond to the general PF pattern (with the exception of the past passive participle constraint on prefixed PF_2 verbs), addition of a prefix to an IMP_2 verb suddenly results in a different aspectual status of the derived verb which becomes PF_4, and its place in the aspectual opposition is taken up by a prefixed *-yva-* verb (just like in the case of the verb *brosat'* 'throw').

Another commonly shared feature for the verbs in Tables 4 – 6 is that, from the point of view of morphology, the future tense of all PF verbs, both base and derived, is the present tense form. It is very clearly observed in the case of the verbs *tolkat'* IMP_2 'push' / *vytolkat'* PF_4 'push out': the present tense form *tolkajet* becomes the future tense *vytolkajet*.

Table 4. The grammatical paradigm for the IMP_1/IMP_2 opposition

	Base Verbs ('LEAD')	+ Spatial Prefix	→	Derived Verbs		
Morphological Type Infinitive	**IMP_1** *vesti*	**IMP_2** *vodit'*	**IMP_3** †/* *vazhivat'*	**PF_3** *vyvesti*	**IMP_{2der}** *vyvodit'*	**IMP_{3der}** ◊ *vyvazhivat'*
Tense						
Past	*vjol*	*vodil*	† *vazhival*	*vyvel*	*vyvodil*	*vyvazhival*
Present	*vedjot*	*vodit*	* *vazhivajet*	------	*vyvodit*	*vyvazhivajet*
Future	budet *vesti*	budet *vodit'*	* budet *vazhivat'*	*vyvedet*	budet *vyvodit'*	budet *vyvazhivat'*
			PARTICIPLE			
Present Active	*vedushchij*	*vodjashchij*	* *vazhivajushchij*	------	*vyvodjashchij*	*vyvazhivajushchij*
Present Passive	*vedomyj*	*vodimyj*	* *vazhivajemyj*	------	*vyvodimyj*	*vyvazhivajemyj*
Past Active	*vedshij*	*vodivshij*	* *vazhivavshij*	*vyvedshij*	*vyvodivshij*	*vyvazhivavshij*
Past Passive	**veden*	†*vozhen*	* *vazhivan*	*vyveden*	* *vyvozhen*	* *vyvazhivan*
			VERBAL ADEVERB			
Present	*vedja*	*vodja*	* *vazhivaja*	------	*vyvodja*	*vyvazhivaja*
Past	* *vev*	* *vodiv*	* *vazhivav*	*vyvev*	* *vyvodiv*	* *vyvazhivav*

◊ - metaphorized, † - obsolete, * - reconstructed

Table 5. The grammatical paradigm for the PF_1 /IMP_2 opposition

Base Verbs ('GIVE') + Spatial Prefix → Derived Verbs				
Morphological Type Infinitive	**PF_1** *dat'*	**IMP_2** *davat'*	**PF_3** *vydat'*	**IMP_{2der}** *vydavat'*
Tense				
Past	*dal*	*daval*	*vydal*	*vydaval*
Present	------	*dajot*	------	*vydajot*
Future	*dast*	budet *davat'*	*vydast*	budet *vydavat'*
		PARTICIPLE		
Present Active	------	*dajushchij*	------	*vydajushchij*
Present Passive	------	*davajemyj*	------	*vydavajemyj*
Past Active	*davshij*	*davavshij*	*vydavshij*	*vydavavshij*
Past Passive	*dan*	**davan*	*vydan*	* *vydavan*
		VERBAL ADEVERB		
Present	------	*davaja*	------	*vydavaja*
Past	*dav*	* *davav*	*vydav*	* *vydavav*
* - reconstructed				

Table 6. The grammatical paradigm for the PF_2/IMP_2 opposition

	Base Verbs ('PUSH')	+ Spatial Prefix	→	Derived Verbs		
Morphological Type Infinitive	**PF_2** *tolknut'*	**IMP_2** *tolkat'*	**IMP_3** † *talkivat'*	**PF_{2der}** *vytolknut'*	**PF_4** *vytolkat'*	**IMP_{3der}** *vytalkivat'*
Tense						
Past	*tolknul*	*tolkal*	† *talkival*	*vytolknul*	*vytolkal*	*vytalkival*
Present	------	*tolkajet*	* *talkivajet*	------	------	*vytalkivajet*
Future	*tolknjot*	budet *tolkat'*	* budet *talkivat'*	*vytolknet*	*vytolkajet*	budet *vytalkivat'*
			PARTICIPLE			
Present Active	------	*tolkajushchij*	* *talkivajushchij*	------	------	*vytalkivajushchij*
Present Passive	------	*tolkajemyj*	* *talkivajemyj*	------	------	*vytalkivajemyj*
Past Active	*tolknuvshij*	*tolkavshij*	* *talkivavshij*	*vytolknuvshij*	*vytolkavshij*	*vytalkivavshij*
Past Passive	* *tolknut*	†*tolkan*	* *talkivan*	*vytolknut*	* *vytolkan*	* *vytalkivan*
			VERBAL ADEVERB			
Present	------	*tolkaja*	* *talkivaja*	------	* *vytolkaja*	*vytalkivaja*
Past	*tolknuv*	* *tolkav*	* *talkivav*	*vytolknuv*	*vytolkav*	* *vytalkivav*

† - obsolete, * - reconstructed

The above observations may be summed up as follows:

1. There are lexically equivalent pairs of verbs of the *vesti/vodit'* type traditionally treated as belonging to the same imperfective aspect.

2. Upon addition of a spatial prefix, IMP_1 verbs become perfective, while IMP_2 verbs with prefixes do not change their aspectual status.[22]

3. The perfective aspect is incompatible with present tense.

4. Morphologically, the future tense of perfective verbs is a present tense.

5. All perfective verbs, both base and derived (with the exception of suffixed PF_2 verbs which are subject to certain constraints), yield past passive participles with a resultative meaning. IMP_2 past participles, both base and derived, gradually fall out of use, and when used possess a factual meaning.

Looking for possible explanations to these interesting phenomena, let us now turn to the functional characteristics of aspectually paired verbs in discourse.

6.2.4. The Syntax of Aspect

As has been shown by Maria Glovinskaya (1982), there is a certain relationship between the aspect of the verb and the general meaning of the context in which it is used, and an analysis of verbal aspect must take into account the figure of *observer*. Similar suggestions have been made by other authors as well (Paducheva 1986; Koshelev 1988). Moreover, the Russian term *vid* used to refer to the respective grammatical category, is related to the verb *videt'* 'see' with its etymological roots traced to the Lat. *videre* and Gk εἶδος 'that which is seen'. However, this relationship between the grammatical meaning of the Russian verb and the figure of observer has not been given sufficient attention. In keeping with the 'conduit metaphor', when verbal communication is seen as transfer of thoughts from the speaker's head to the listener's head, traditional linguistic theory lays heavy emphasis on the role of the speaker in explaining linguistic phenomena at various levels of complexity (the so-called concept of egocentricity of language). However, as I have shown elsewhere (Kravchenko 1992, 1993, 1995), the primary point of reference for indexical phenomena both in lexicon and grammar is not the speaker but the observer, or the *speaking (languaging) observer*. If we view natural language grammar as a sign system for representing cognitive experience through conceptualization and categorization where meaning is sought in the

---

[22] It must be noted, however, that some IMP_2 verbs may become perfective with an addition of a prefix — as, for example, in *tolkat'* 'push' (IMP_2)— *vytolkat'* 'push out' (PF_4). The difference in meaning in this case is that of the manner in which the action is performed: *vytolknut'* refers to a one-time action, while *vytolkat'* refers to a series of successive actions viewed holistically.

realm of cognitive processing (Langacker 1991), then any insightful analysis of linguistic categories is impossible without taking into account the primary cognitive dimension of speech, that is, the observer, as 'everything said is said by an observer to another observer'. So let us now take a look at the functional characteristics of aspectually paired verbs in two types of context of which one contains explicit indication to the observer as the source of information about the event, while the other is neutral in this respect.

PF_1/IMP_2 and PF_2/IMP_2 Verbs

Consider the following examples:

(37) *Posmotri, on chto-to brosil/*brosal v urnu.*
Look, he something throw PF_1/*IMP_2 PT into trash can A
'Look, he dropped something into the trashcan.'

(38) *Kazhdyje polchasa on *brosil/brosal v kamin novoje poleno.*
Every half hour he throw *PF_1/IMP_2 PT into fireplace A new log A
'Every half hour he tossed a new log in the fireplace.'

(39) *Chelovek, chto dal/*daval vam sejchas prikurit', — kto on?*
Man that give *PF_1/IMP_2 PT you now light up INF who he?
'The man who just gave you a light — who is he?'

(40) *Otchim, xot' i *dal/daval mne rubl' po voskresenjam,*
Stepfather though and give *PF_1/IMP_2 PT me ruble on Sundays
menja ne ljubil.
me not love IMP_2 PT
'My stepfather, even though he would give me a ruble on Sundays, didn't love me.'

In (37) and (39) the speaker describes directly observed events (direct observation is explicated contextually by the imperative *Posmotri* in (37) and by the indexical *sejchas* in (39)), so IMP_2 cannot be used. In (38) the verb phrase expresses the event as a one-time action, but the adverbial phrase *every half hour* indicates that the proposition is not based on some concrete information as the result of an observational event (that is, something that receives a unique spatio-temporal characteristic), but on some generalized knowledge as a result of multiple observations. In other words, the production of utterance (38) does not depend on what can or could be observed at a given moment, but on what is known; so, PF_1 cannot be used. In (40) the event ("the giving of a ruble") is viewed as part of an indefinite sequence of similar events: the speaker tells about something he knows because of repeated multiple experience, and the use of PF_1 is precluded.

In the following example the speaker refers to an event that just took place and the result of which can be observed, so only PF_2 is allowed:

(41) *Ax, izvinitje, ja ne xotel vas tolknut'/*tolkat'!*

Oops, excuse, I not want IMP_2 PT you push PF_2/ *IMP_2 INF

'Oops, sorry, I didn't mean to push you!'

However, in (42) where the speaker is undoubtedly the observer, only IMP_2 is allowed:

(42) *Perestan' menja *tolknut'/tolkat'!*

Stop me push *PF_2/IMP_2 INF

'Stop pushing me!'

How can this peculiarity be explained? In the above examples, the PF_1 and PF_2 verbs refer to actions which are categorized as such on condition that a certain change took place and was observed before the moment of utterance in which the corresponding verb is used (hence the concepts of "boundedness" and "totality"). The meaning of such verbs as *begin, continue, stop* — regardless of their aspectual class — is incompatible with PF since it is not possible to begin, continue, or stop doing something that is absent in the observed and described situation. Sentence (42) describes a situation in which either someone tries to physically move me from the place where I am, or keeps trying to make me move by repeatedly applying physical pressure. In the first case, the completion of the action is not followed by its expected natural outcome (change of my position in space), and the PF verb cannot be used. In the second case, reference is to a series of actions of pushing, whereas a typical PF verb refers to a one-time action. As a consequence, in both the cases only the IMP verb may be used.

Sentences with a negative pronoun as the subject, such as *nobody*, do not allow the use of PF when *nobody* has the generalized meaning "not one of all the existing persons", cf.:

(43) *Kto tebja tolknul, synok? — Nikto menja ne *tolknul/tolkal.*

Who you push PF_2 PT sonny? — Nobody me not push *PF_2/IMP_2 PT

'Who pushed you, sonny? — Nobody pushed me.'

(44) *Kto tebja sjuda pustil? — Nikto menja ne puskal.*

Who you here let in PF_1 PT? — Nobody me not let in IMP_2 PT

Ja sam sebja pustil.

I myself me let in PF_1 PT

'Who let you in here? — Nobody let me in. I let myself in.'

The negative subject *nobody* implies that the action referred to did not really take place. If an action did not take place, it could not be observed; therefore, the entire event (or non-event, to be precise) is presented by the speaking observer as inferred knowledge rather than the result of live observation (it is, of course, possible for the statement itself to be false). The question *Who pushed you, sonny?* is asked because the speaker observes something that he interprets as the result of "pushing" as an action, with *sonny* as its object; in this case only the PF may be used. Conversely, the boy who is the speaker now uses the IMP which does not imply observability of the event. In (44) the context of both utterances is similar to that in (43), but in *I let myself in* PF is used just like in the opening utterance *Who let you in here?* because the speaker makes a statement about something that he himself observes. Indefinite reference characteristic of *nobody* in such uses ("not one of a given set of persons") requires a PF verb, as in (45):

(45) *Chego on plachet? Nikto jego dazhe ne tolknul/*tolkal.*

What G he cry IMP_2 PR. Nobody him even not push PF_2/*IMP_2 PT

'Why is he crying? Nobody even as much as jostled him.'

As a rule, PF_1 and PF_2 verbs have one-time actions as their referents, and it affects the interpretation of utterances with such verbs, as in the following examples:

(46) a. *Ona rodila devochek.*

She give birth PF_1 PT girls

'She delivered girls.'

b. *Ona rozhala devochek.*

She give birth IMP$_2$ PT girls

'She kept bearing girls.'

(47) a. *On djornul za verjovki.*

He pull PF$_2$ PT on ropes

'He pulled on the ropes'

b. *On djorgal za verevki.*

He pull IMP$_2$ PT on ropes

'He kept pulling on the ropes'

In (46a) the act of giving birth is viewed as a one-time event whereby twins (triplets, etc.) are delivered. In (46b) the act of giving birth is viewed as an unspecified number of events whereby an unspecified number of female babies are delivered. Likewise, in (47a) *ropes* is a multiple object of a one-time action, whereas (47b) is about an unspecified number of actions each with an unspecified number of objects.

IMP$_1$/ IMP$_2$ Verbs

Consider the following examples:

(48) *Smotri, kuda eto ona jego vedjot/*vodit?*

Look where this she him lead IMP$_1$/*IMP$_2$ PR

'Look, where is she taking him?'

(49) *V kontse ulitsy pokazalsja slon, kotorogo veli/*vodili*

In end L street G showed REFL PF$_3$ elephant which A lead IMP$_1$/*IMP$_2$ PT

dva fakira.

two fakirs

'At the end of the street appeared an elephant lead by two fakirs.'

The context of both utterances includes explicit reference to the observer (the verbs *smotri* and *pokazalsja*), and only IMP_1 is permissible. In the following example no reference to the observer is made, and thus the use of IMP_1 is precluded:

(50) *Kto tebja obychno *vedjot/vodit v shkolu?*

Who you A usually lead *IMP_1/IMP_2 PR in school A?

'Who usually takes you to school?'

Or consider another interesting example, which is an anecdotal answer to the question asked by a biology teacher in a classroom.

(51) TEACHER: *Chto ty znajesh o zhivotnom mirje?* ('What do you know about animal life?')

STUDENT: — *Ptitsy *letjat/letajut, zveri *begut/begajut, ryby*

Birds fly *IMP_1/IMP_2 PR, beasts run *IMP_1/IMP_2 PR, fish

**plyvut/plavajut, zmei *polzut/polzajut.*

swim *IMP_1/IMP_2 PR, snakes crawl *IMP_1/IMP_2 PR

'Birds *are flying/fly, beasts *are running/run,
fish *are swimming/swim, snakes *are crawling/crawl.'

When answering the question in (51), the speaker chooses IMP_2 verbs prompted by the verb *know* used in the question, that is, he makes a statement on the grounds of what he knows about such objects as birds, beasts, etc. Were he to use IMP_1 verbs, his answer would be communicatively inappropriate because he would not be saying something about what he knew, but about what he was observing at the moment of utterance. Every time I presented this example to different students in a classroom choosing the IMP_1 verbs, the reaction was always similar — smiles, giggles, questioning looks and spontaneous comments on the weirdness of what they heard — because, as some would say, 'there were no birds or fish in the classroom'. Likewise, if someone — let us say, in a city park — approached me and said with an assertive intonation:

(52) *Zmei polzajut.*

Snakes crawl IMP_2 PR

'Snakes crawl.'

my most probable reaction would be "So what?" (meaning something like "It's common knowledge", or "Everybody knows that, so why are you telling me?"). However, if this someone were to say:

(53) *Zmei polzut.*

Snakes crawl IMP$_1$ PR

'Snakes are crawling.'

my most likely response would be to start looking around in apprehension, asking, "Where?"

6.2.5. The Meaning of Aspect

Examples illustrating the difference in meaning between aspectually paired verbs such as given in the previous section could be provided *ad infinitum*. This distinction consistently reflects the contrast between two types of knowledge : phenomenological, which is based on individual empirical experience, and structural, which is based on abstraction. In other words, the grammatical term *vid* (literally meaning "view") unambiguously specifies the distinctive feature of aspectual contrasts which we may call the 'cognitive status' of the event expressed by the verb: observed events are contrasted to events reference to which does not imply the observer. Consequently, paired verbs of the *idti/xodit'* ('go') type traditionally labeled as IMPERFECTIVE, constitute *bona fide* aspectual oppositions just as PF$_1$/IMP$_2$ and PF$_2$/IMP$_2$ verbs do.

The difference between the cognitive statuses of paired verbs was intuitively felt by Potebnja (compare his "concrete" vs. "habitual" events) and by Jakobson ("determinate" vs. "indeterminate" aspect). Other explanations of the semantic difference between the paired verbs ("duration" vs. "frequency", or "directedness" vs. "non-directedness") also stem from the cognitive semantics of the verb: motion in space as an event categorized by an aspectually marked IMP$_1$ verb is, in a prototypical situation, present (= exists) at the moment of utterance, it unfolds even as the speaker is describing it — hence the idea of duration associated with such event. When a moving object is being observed, the idea of directedness of its motion is always present as the observer's gaze follows the object's trajectory. By contrast, an aspectually unmarked IMP$_2$ verb expresses the knowledge the speaker has about the event when the source of this knowledge may be either a one-time observation as in (54), or multiple observations as in (55). Note that in (54) the speaker is not necessarily the observer, he may have learned about the event he describes by hearsay:

(54) *Vchera Petja xodil v kino.*

Yesterday Pete go IMP$_2$ PT in movies

'Yesterday Pete went to the movies.'

(55) *Petja xodil po komnate*

Pete go IMP$_2$ PT along room D

'Pete was pacing the floor.'

The IMP$_2$ verb in (55) refers to a process that consists of a number of phases ("to" and "fro") that can be singled out empirically; in a prototypical situation of observation, reference to each single phase would be made with the help of the IMP$_1$ verb *idti,* as in (56):

(56) *Petja idjot po komnate.*

Pete go IMP$_1$ PR along room D

'Pete is moving across the room.'

A sequence of observations registers the recursive character of such phases, which accounts for the feature "frequency" ascribed to IMP$_2$ verbs in traditional grammars.

As a grammatically consistent analysis shows, the term *vid* unambiguously specifies the distinctive feature of aspectual contrasts in Russian, namely, the cognitive status of the event expressed by the verb: observed events are categorized in the form of aspectually marked verbs, whereas aspectually unmarked verbs categorize events without reference to observation. Consequently, the grammatical meaning of aspect has very little to do with "boundedness" or "totality" as the so-called semantic invariant but must be defined as *indication to the source of information about the event* which can be *definite* (based on observation), or *indefinite* (based on the speaker's knowledge whose source is not specified). Thus, the primary aspectual opposition is constituted by forms contrasted on the cognitive-semantic feature "definite/indefinite"; we should speak of Definite Aspect and Indefinite Aspect as the two aspect classes which include all Russian verbs. Depending on the type of observation categorized in the semantics of the definite aspect verbs they further fall into two types that may be labeled "definite imperfective" (IMP$_1$) and "definite perfective" (various morphological types of PF verbs). In other words, there is a *tripartite aspect system* in Russian. IMP$_2$ verbs which do not have IMP$_1$ counterparts subsume the two aspectual distinctions characteristic of paired verbs of the *vesti* IMP$_1$ – *vodit'* IMP$_2$ ('lead') type. One or the other aspectual meaning ("definite" or "indefinite") is brought out by the context via explicit indication to the observer or the observability of the situation described in a sentence, as in (57):

(57) a. *Ivan igrajet v simfonicheskom orkestre.*

Ivan play IMP PR in symphonic L orchestra L

'Ivan plays in the symphonic orchestra' (i. e., 'this is what I know about Ivan').

b. *Ty slyshish zvuki muzyki? — Eto Ivan igrajet na flejte.*

You hear sounds music G? — This Ivan play IMP PR on flute L

'Can you hear the sounds of music? — That's Ivan playing the flute' (i. e., 'what you hear is Ivan playing the flute').

6.3. Aspect and the Participle

6.3.1. The Morphology of Participles and Constraints on Their Use

As in many other languages, there are two participles in Russian, the active and the passive (see Tables 4-6), each of which has two forms traditionally labeled as 'present tense participle' and 'past tense participle' (Shvedova and Lopatin 1989) and is typically used in the attributive function in a sentence. I will be using these terms as officially acknowledged labels, although, as will become clear from the following discussion, they are conceptually misleading.

Table 7. The forms of the Russian participle

Infinitive	Conjugation Class	Active Participle	Passive Participle
		Present / Past	Present / Past
Transitive			
nesti CARRY	I	*nesushchij / nesshij*	*nesomyj / *nesennyj*
chitat' READ	I	*chitajushchij /chitavshij*	*chitajemyj / chitannyj*
Intransitive			
lezhat' LIE	II	*lezhashchij / lezhavshij*	------- / ------
sidet' SIT	II	*sidjashchij / sidevshij*	------- / ------

Active participles are systemically opposed to passive participles in the case of transitive verbs while they don't have corresponding correlates in the case of intransitive verbs as is shown in Table 7. Both the active participles appear to have the same verbal stem (e. g. *chita-*), seemingly differing in the suffixes that are added: *-(j)ushch-/-(j)ashch-* + the ending *-ij/-aja/-jeje* which marks the participle for masculine (M), feminine (F) or neuter (N) gender for the present participle, and *-(v)sh-* + the ending *-ij/-aja/-jeje* for the past participle. The participle retains the aspectual category of the verb from which it is derived, yet perfective verbs do not yield present participles, which is in keeping with the inability of finite perfective verbs to have present tense forms. Thus, the category of verbal aspect — as it is traditionally understood and interpreted in Russian studies — appears to be intrinsically related to the category of tense insofar as the so-called perfective aspect is concerned (see Table 8).

Table 8. Tense-aspect correlation for perfective verbs

Infinitive	Aspect	Lexical Meaning	Present Tense	Active Participle	
				Present	Past
past'	PF_1	FALL	-------	-------	*pavshchij*
tolknut'	PF_2	PUSH	-------	-------	*tolknuvshij*
vytolkat'	PF_4	PUSH OUT	-------	-------	*vytolkavshij*
vyvesti	PF_3	TAKE OUT	-------	-------	*vyvedshij*
padat'	IMP_2	FALL	*padajet*	*padajushchij*	*padavshchij*
tolkat'	IMP_2	PUSH	*tolkajet*	*tolkajushchij*	*tolkavshij*
vesti	IMP_1	LEAD/TAKE	*vedjot*	*vedushchij*	*vedshij*

Although, morphologically, the opposed pairs of active participles are described in Russian grammars as oppositions based on the temporal distinction PRESENT/PAST, it is nothing but homage to the history of the Russian language. In Old Russian these two participles, being verbal adjectives, could express between themselves temporal distinctions, as one of them was derived from the present tense verb stem and the other from the aorist (Borkovskij and Kuznetsov 1963; Ivanov 1983). This feature is reflected in the impossibility for the past active participle to be used in a sentence/utterance where the main verb is in the present tense, cf.:

(58) a. *Vidish von tu tantsujushchuju/*tantsevavshuju paru?*

See IMP PR 2 S yonder that A dancing IMP PA PR /PT couple A

'Can you see that couple dancing over there?'

b. *Ja ljublju smotret' na igrajushchix /*igravshix shchenjat.*

I like IMP PR look IMP INF on play IMP PA PL PR /*PT puppies A

'I like to watch playing puppies.'

At the same time the so-called present participle can be used in a past tense context set by the main verb, just like its past tense counterpart, cf.:

(59) a. *Ivan podoshol k plachushchemu mal'chiku.*

Ivan come up PF PT to cry IMP PA PR boy D

'Ivan came up to the crying boy.'

b. *Ivan podoshol k plakavshemu mal'chiku.*

Ivan come up PF PT to cry IMP PA PT boy D

'Ivan came up to the crying boy.'

It is cases like these that pose a problem: superficially, there seems to be no functional-semantic difference between the two participle uses illustrated in example (59). Or, to put it differently, regardless of what grammars tell us, active participles in contemporary Russian cannot express temporal distinctions of the kind characteristic of the finite verbs used in a specific grammatical tense such as present or past, since both can be legitimately used in describing a past situation as determined by the main verb. Compare examples (60) and (61):

(60) a. *Ivan obratil vnimanije na stojashchuju za uglom mashinu.*

Ivan turned attention on stand IMP PA PR around corner car A

'Ivan noticed a car standing around the corner.'

b. *Ivan obratil vnimanije na stojavshuju za uglom mashinu.*

Ivan turned attention on stand IMP PA PT around corner car A

'Ivan noticed a car standing around the corner.'

(61) a. *Ivan osmotrelsja. Jego vnimanije privljek ulybajushchijsja*

Ivan looked around. His attention was caught by smile IMP PA PR

molodoj chelovek, stojashchij u okna.

young man stand IMP PA PR by window G

'Ivan looked around. His attention was caught by a smiling young man standing by the window.'

b. *Ivan osmotrelsja. Jego vnimanije privljek ulybavshijsja*

Ivan looked around. His attention was caught by smile IMP PA PT

molodoj chelovek, stojavshij u okna.

young man stand IMP PA PT by window G

'Ivan looked around. His attention was caught by a smiling young man standing by the window.'

This intriguing phenomenon is not accounted for in contemporary Russian grammars. At the same time, in certain past contexts only one of the two forms can be used while the other is constrained, as in (62):

(62) *Po slovam ochevidtsev avarii, chelovek, sidevshii/*sidjashchii*

To words D eye-witnesses G accident G man N sit IMP PA PT / *IMP PA PR

rjadom s voditelem, umer srazu.

next to driver INS die PF PT at once

'According to the testimony of the witnesses of the accident, the man *who had been sitting* / **sitting* next to the driver died instantly.'

Data of this kind allow to speak of the *functional asymmetry* of the Russian active participles which cannot be explained by relating it to the alleged difference between their respective temporal reference domains implied by the labels 'present tense participle' and 'past tense participle'. On the one hand, the behavior of perfective verbs in regard to participle formation (as illustrated in Table 8) suggests that the semantics of the present participle has something to do with the semantics of aspect. On the other hand, since perfective meaning is incompatible with the meaning of the present participle to the degree that it precludes formation of such participle from perfective stems, it becomes obvious that the meaning of the present tense participle has little to do with temporal distinctions as specified by the system of finite verbs. This poses the question, "If the meaning and function of the present tense participle are not determined by temporal reference, how can they be explained?"

6.3.2. The Nature of the Functional Asymmetry of Participles

The fact that functional asymmetry of the Russian active participles has not been given due attention in theoretical grammar research, not to mention practical (both descriptive and prescriptive) grammar guides be they for native speakers of Russian or for students of Russian as a foreign language, especially against the background of alarmingly rapid growth of functional illiteracy in contemporary society, discussed in Chapter 4, has led to an increasing indiscrimination between the uses of the two participles in contexts where only one form of the participle is admissible. Such indiscrimination affects the ease with which texts are interpreted; particularly, it obscures the distinction between *different points of view* that may be, and often are, expressed when reporting events, and which cue the reader's understanding. Consider an example from a newspaper article about a convention of Russian cinematographers (published in *Kommersant*, with circulation over 120,000 copies) where the use of the present participle creates a curious effect bordering on anomalousness, when the situation described in the sentence is perceived as if viewed by the participating person (*Elena Uvarova*), while in fact it is just a reporter's account of what the reporter saw (63a). To rid the sentence of this strong tinge of anomalousness, the past tense participle (*prisutstvovavshej*) must be used instead of the present (63b):

(63) a. ? *Vozmozhno, eto soobshchenije perepolnilo chashu terpenija*

Perhaps this information overfill PF PT cup A patience G

prisutstvujushchej v zale Eleny Uvarovoj.

be present IMP PA PR G in hall L Elena G Uvarova G

? 'Perhaps this information was the last drop that made the cup run over for Elena Uvarova being (present) in the conference room'.

b. *Vozmozhno, eto soobshchenije perepolnilo chashu terpenija*

prisutstvovavshej v zale Eleny Uvarovoj.

'Perhaps this information was the last drop that made the cup run over for Elena Uvarova who was (present) in the conference room'.

Alarmingly, such incompetent use of participles may be observed not only in the media but in the literary works by some popular contemporary Russian writers as well. As a result of such misuse of active participles, the perceptual perspective on the situation described in the sentence gets distorted and two different points of view are confused, one of the Speaker and the other of the Observer. Obviously, the conceptual distinction between the two forms of participles, categorized as their grammatical meaning, is not one of time as suggested by the traditional labels "present" and "past", but of the source of information about the described event or situation, which is to say that it is an *aspectual distinction*. That it is, indeed, so is illustrated by the fact that active participles which belong to different aspectual classes consistently display the same kind of constrained relationship in similar or analogous contexts.

As was shown in the previous section, common to IMP_1, PF_1, and PF_2 verbs is that they are all marked for the cognitive parameter 'observer' which is at the phenomenological core of the grammatical category of aspect. Naturally, PF_3 verbs derived from IMP_1 verbs also retain this feature. However, even though spatial prefixes in modern Russian have become the dominant means of changing the aspectual class of the verb from so-called 'imperfective' to 'perfective', we'll be looking only at the *basic* aspect types shown in Table 9, since prefixed verbs are a later development in the Russian morphological system. (Although passive participles are not analyzed here, they have been included in the table to give a fuller picture. It is important to note, however, that the IMP_1 and the IMP_2 present passive participles are subject to the same kind of syntactic constraints as in the case of active participles, which are described and analyzed below).

On the one hand, as has already been mentioned, the so-called perfective verbs (both base and derived) do not have a present tense or corresponding present tense participles. On the other hand, lexically paired verbs of locomotion which belong to different aspectual classes (definite IMP_1 vs. indefinite IMP_2) yield present tense active participles which consistently display constraints on their use in certain contexts:

(64) a. *On pritselilsja v letjashchuju ptitsu.*
He took aim at fly DEF IMP_1 PR PA F S A bird A
'He took aim at the flying bird.'

b. * *On pritselilsja v letajushchuju ptitsu.*
He took aim at fly INDEF IMP_2 PR PA F S A bird A
'He took aim at the flying bird.'

Moreover, past tense active participles derived from indefinite imperfective stems of paired locomotion verbs display the same kind of constraint, thus differing

Table 9. Aspect types and the participle

ASPECTUAL CLASS / MORPHOLOGICAL FORMS	DEFINITE (marked for observer)		INDEFINITE (unmarked for observer)	LEXICAL MEANING
	ASPECT TYPE (cognitive parameter)			
	PF_1 perceptual pastness	**IMP_1** perceptual presentness	**IMP_2** perceptually unspecified	
INFINITIVE				
Transitive	*brosit'*	------	*brosat'*	throw
	------	*katit'*	*katat'*	roll
Intransitive	------	*letet'*	*letat'*	fly
ACTIVE PARTICIPLE				
Present	------	------	*brosajushchij*	throwing
	------	*katjashchij*	*katajushchij*	rolling
	------	*letjashchij*	*letajushchij*	flying
Past	*brosivshij*	------	*brosavshij*	throwing
	------	*kativshij*	*katavshij*	rolling
	------	*letevshij*	*letavshij*	flying
PASSIVE PARTICIPLE				
Present	------	------	*brosajemyj*	being thrown
	------	*katimyj*	*katajemyj*	being rolled
	------	------	------	*being flown
Past	*broshennyj*	------	------	thrown
	------	------	*katannyj*	rolled
	------	------	------	*being flown

from the past tense participles of IMP_2 verbs of the general stock (that is, the majority of IMP_2 lexically unpaired verbs) which are not constrained for use in a past context as illustrated in examples (59b), (60b) and (61b) above. Cf.:

(65) a. *On pritselilsja v letjevshuju ptitsu.*
He took aim at fly DEF IMP_1 PT PA F S A bird A
'He took aim at the flying bird.'

b. * *On pritselilsja v letavshuju ptitsu.*
He took aim at fly INDEF IMP2 PT PA F S A bird A
'He took aim at the flying bird.'

Contexts which display constraints on the use of certain types of participles contain, typically, indication to the *observability* of the state of affairs or events described; thus, in example (62) it is the nominal phrase *ochevidtsy avarii* 'eye-witnesses of the accident', in (64) and (65) it is the verb *pritselilsja* 'took aim'. Interestingly, similar contexts do not impose any constraints on the use of participles formed by unpaired indefinite aspect verbs, as becomes obvious from examples such as (60) and (61). For that matter, such participles are usually sententially unconstrained when the main verb is in the past tense, that is, within an isolated sentence both the past and the present tense participle would be grammatically correct, as in (59), except in the cases such as illustrated in (63), when propositional modality (the modal marker *perhaps*) unambiguously identifies the speaker as the observer and, therefore, the source of information.

Now if we address the question asked earlier, "If the meaning and function of the present tense participle are not determined by temporal reference, how can they be explained?", it should have become clear by now that the systemic grammatical meaning of the present tense active participle is determined by the aspectual type of the verb stem. Because imperfective definite aspect verbs are marked for observer as the source of information, the use of corresponding present participles implies reference to a situation-event as something in the focus of current perception. Therefore, while interpreting any text fragment which contains a definite imperfective present participle, the cognitive parameter 'observer' *must* be ascribed to a contextually identifiable human subject. This is, so to speak, the tacit knowledge every educated native speaker of Russian possesses. 'Educated' here means not so much 'having received formal education', but rather 'well-read in one's mother-tongue literature', and education thus understood is viewed as a necessary prerequisite for functional literacy in the world where "texts rule" (see Chapter 4 for discussion).

However, such identification relies crucially on the (usually intuitive) ability of the interpreter to correlate the situation-event, morphologically categorized by the participle as 'currently perceived' (which is the grammatical-indexical meaning of definite aspect) with the split temporal perspective sustained by the use of tensed finite verbs on the one hand, and the indexical participle, on the other. But what does this actually mean? It means that the speaker, who is also always an observer by default (remember that everything said is said by an observer to another observer), when describing a past situation-event projects it into a temporal domain which does not overlap with the speakers domain of current perceptually registered interactions with the environment, that is, his present. By using in this context an indexical participle (for example, *letjashchaja* 'flying' in (64a)), the speaker indicates that the described situation-event (bird-hunting in this example) is viewed from a vantage point located in the past, and it is viewed as something perceptually present, thus bringing in the figure of observer. Since *he* in *He took aim at the flying bird* does not refer to the speaker, its referent is identified as the actual observer of the locomotive event *flying.* Now it becomes clear why example (64b), in contrast to (64a), is semantically anomalous: the verb *pritselilsja* 'took aim' as a lexical identifier of the observer in the described situation does not agree with the present tense indefinite aspect participle *letajushchaja* 'flying' which is unmarked for the observer as the source of information. The past tense definite aspect participle *letevshaja* and the past tense indefinite aspect participle *letavshaja* in example (65) display similar constraints for the same reason.

If we now ask, "What is the difference in meaning between examples (64a) and (65a)?", the answer will be that in (64a) we deal with two conjoint points of view expressed in linguistic form within one sentence, one of the speaker and the other of the hunter who is not the speaker, whereas in (65a) it is only the speaker's point of view because the past participle *letevshaja* sustains a singular temporal perspective by referring to a past situation-event (the historical consequence of its decent from the aorist). Similarly, in examples (59) through (61) we are dealing with the same phenomenon, when an observer other than the speaker is identified in the first sentence in each pair, while in the second sentence the sole source of information is the speaker.

To sum up, the functional peculiarities of the present and past active participles can be explained by relating their referential features to the cognitive parameter "source of information about the event" which can be either the speaker (author of the sentence/utterance), or the observer (the actual witness/participant of the described situation-event). Thus we can speak of two pragmatic axes, those of the Speaker and the Observer, which serve as the lines of reference for the temporal features possessed by the main and background events, accounting for the "different-point-of-view" effect. Violation of cognitive agreement between these lines of reference results in semantic anomalousness as illustrated in example (63).

Well, now, what about examples such as (62)? Indeed, there may seem to be a contradiction here between the inferences just made and the actual nature of constraint on the use of the present participle *sidjashchij* 'sitting'. If the participle refers to something currently observed and the observer is unambiguously identified (it is 'eye-witnesses of the accident'), why cannot the participle be used, after all? What we are looking at here is another interesting aspect of the cognitive semantics of the present participle — its epistemic agreement with the object it characterizes. *Sit* (Russian *sidet'*) is an anthropomorphic posture verb used, alongside with *stand* and *lie*, to describe one of the three basic functional postures of humans: standing, sitting, and lying. Elsewhere (Kravchenko 2001b) I have referred to these verbs as 'basic existential predicates' which categorize certain cognitive structures incorporated in the corresponding concepts associated with these lexical items. The manner of categorization depends on how humans perceive different objects which are classified on such criteria as size and shape comparability, mobility, the accepted norm, as well as morphological structure *in the case of living creatures.* As a predicate, *stand* can be meaningfully used in the case of human subjects only on condition that the person in question is *alive.* Dead bodies do not typically stand or sit, although they can and do lie, of course. The noun *chelovek* 'man' (that is, 'live human being') in example (62) refers to a past temporal domain from the point of view of the witnesses (who are assigned both the roles of speakers and observers), because at the time when they bear testimony their perceptual experience of the accident is not concurrent with their present perceptual experience (that is, at the moment of speaking). In other words, at the time they describe the accident the person they refer to is no longer alive, and they could hardly have seen him alive before the accident. The use of a present tense participle *sidjashchij* 'sitting' in this example would unambiguously convey the meaning 'when they saw him after the wreck he was alive' which, naturally, would be in contradiction with the assertion 'he died at once'. Notice that exactly the same type of constraint imposed by the semantics of the present participle is observed in the English translation *'According to the testimony of the witnesses of the accident, the man* who had been sitting / *sitting *next to the driver, died instantly'.*

6.4. Aspect in English

6.4.1. Divorcing Aspect from Tense

In contemporary linguistic literature tense, along with the category of verbal aspect, has been one of the most controversial issues. This controversy, I believe, has been caused, to a large degree, by misunderstanding and, as a result, misinterpretation of the genuine grammatical meaning of both categories. Rather than trying to explain the meaning of a particular grammatical category through an application of some arbitrary semantic procedure rooted in the formal methods of logic, linguists should look for simple cognitive mechanisms — many of which are perceptually grounded — that lead to particular conceptualizations of the multifarious aspects of the so-called 'real world'. The above mentioned controversy is largely rooted in the fact that in many languages of the world the number of verb forms referred to as "tenses" is more than three, so these forms do not fit very well in the conventional triadic schema known as the "time axis" shown in Figure 7.

According to the long established tradition, tense is believed to be a grammatical category that correlates most directly with distinctions of time in terms of present, past, and future. This means that the grammatical meaning of tense cannot be fully understood and interpreted without the nature of time being considered. Therefore, the concepts of present, past, and future should be analyzed from the point of view of their cognitive content, aiming to answer the number one question, "What is "present" ("past", "future")?" The conceptual paradigm for treating tense as a grammatical category depends critically on how this question is answered.

```
PAST                          PRESENT                          FUTURE
∞ ---------------------------------x----------------------------------> ∞
```

Figure 7. The time axis

The term "tense" traditionally refers to finite forms of the verb. There are twelve such forms in English in the active voice, and grammarians usually speak of twelve tenses. At the same time, it is generally understood that there can be no more than three tenses in any given language — at least, the conceptual model of time in any existing language does not allow room for a fourth element different from PRESENT, PAST, and FUTURE. Thus, the term "tense" in its current grammatical usage is vague and imprecise since the finite forms are nothing but a conglomerate of several grammatical categories that include tense, mood, and aspect (Dahl 1985).

To understand how grammatical tense reflects temporal distinctions, we must look at the origin of time as a *psychological phenomenon, rooted in perceptual experience*. A good clue is given by etymological analysis of corresponding terms, such as *present*, for example. Contrary to the widespread belief that many metaphors are semantically "dead", I don't really think that there are such things as dead metaphors. Moreover, it is often "dead" metaphors that hold the key to the door in many theoretically controversial cases, especially in grammar. For example, the English word *presence* may be traced from the Lat. *præsens*, which consists of the prefix *præ-* 'in front of, before', and the root *-sens*, which is traced to the noun *sensus* 'sense'[23]. So, the meaning of *præsens* is 'that which is before the

23 Although, traditionally, *-sens* is interpreted as the present participle of *esse* (Skeat 1956), there are sufficient reasons to question the validity of this assumption. Following Emile Benveniste's (1971) insightful analysis of the meaning and usage of *praesens* and other words with the same root in Latin, I have argued that the exact meaning of *praesens* shows that *-sens* has very little to do with *esse*. I believe it was the vise-like grip of tradition that prevented Benveniste from

senses', i. e. 'perceived with the senses'. This meaning is preserved in the adjective *present* in such uses as *All are present.* Correspondingly, *absent* means 'that which is not before the senses', as in (66):

(66) *Does anyone know why Mr. Jones is absent from work today?*

The primary meaning of the correlate term *past* is 'beyond in position, father than', i.e., originally *past* meant (and still does) 'that which is gone by, no longer current'. We know that something is gone by when we cannot observe it any longer, so *past* refers to that domain of our experience which is not associated with current observation (is "absent") but is retained in memory. The division of our world experience into two domains ("that which is currently observed" and "that which is no longer observed") serves as a phenomenological background on which our conception of time is based (Kravchenko 1996). It also explains why in many languages of the world — including English — the grammatical category of tense is morphologically represented by a binary opposition such as "past/non-past" or "past/ present": the languaging observer categorizes all his phenomenological experience by dividing it into "everything currently observed" and "everything that is no longer observed". There is no future time in the 'real world', the concept of future has to do with our experience-based prognoses about the possible flow of events, that is, future time is the domain of predicted experience, and in this it differs essentially from present and past time.

The concept of future in many languages emerged much later than other temporal concepts, at the time when their grammars had been formed, so the already existing resources had to be drawn upon for the expression of future events. For example, in English a construction consisting of the modal verb *will* and an infinitive is used to express futurity, and in Russian it is the verb *be* in the future tense followed by an infinitive:

(67) *budu delat'* '(I) will do' = *byt'* FUT 1 S + *do* INF

Futurity in Russian is also expressed by the present tense stem of the verb with a spatial prefix:

(68) *sdelaju* '(I) will do' = *s-* 'with' + delaju 'do PR 1 S'

That the English *will* as an auxiliary in the future tenses retains its lexical (modal) meaning affects its functional (grammatical) properties, resulting in what is

making the obvious conclusion himself, although he had all the facts correctly. The argument for the meaning of the Latin *praesens* is given in Kravchenko (1992) and Kravchenko (1993).

known as the Future-in-the-Past Tense. Although the term itself defies tense logic based on the monolinear model of time as shown in Figure 7, the expression *future in the past* is not self-contradictory as long as a different kind of logic is used when grammatical concepts are viewed as reflecting categorized perceptual experience. The form *would* (past tense of *will*), as in (69a), is used to express futurity not from the speaker's viewpoint (i. e., from the point of view of the person who actually utters something like (69a)), but from the point of view of the person the speaker is talking about (i. e., *he*). In other words, *he would do it* in the above example refers to the future of *he,* which is nothing but the past for the speaker. The meaning of such a sentence is as follows: 'At a certain moment in the past he told me about his intention to do it later, but I am not sure whether he has done it or not, and if not, I am not sure he is going to, anyway'. By the same token, if the speaker has strong reasons to believe that the promise will be kept, the other person's future extends into the speaker's future (shared prognosticated experience), thus making (69b) perfectly grammatical:

(69) a. *He told me he would do it.*

b. *He told me he will do it.*

The meaning of (69b) may be interpreted as follows: 'At a certain moment in the past he told me about his intention to do it later, and although he has not done it yet, I believe he is going to, anyway'. A very similar phenomenon, which may be put down to the perceptual-cognitive difference between the concepts SPEAKER and OBSERVER, is found in Russian as well, as is seen from examples (70a) and (70b):

(70) a. *On osmotrelsja. Raboty predstojalo mnogo.*

He look-around PT PF. Work G be-in-store PT IMP much.

'He looked around. There was a lot of work to be done here.'

(70) b. *On osmotrelsja i uvidel, chto raboty predstoit mnogo.*

He look-around PT PF and see PT PF that work G be-in-store PR IMP much.

'He looked around and saw that there *is* a lot of work to be done.'

In (70a) the past tense is used in both the sentences as an account of the situation is given from point of view of the speaker. In (70b) the cognitive verb *see* relates the proposition expressed in the introduced clause to the observer (*he*) who is a person other than the speaker. As may be seen from examples like these, the concept of time as a knowledge structure is characterized by intrinsic perceptual groundedness; this allows us to speak of the cognitive underpinning in the meaning of the grammatical category of tense, which is *indication of the relationship between the described event and the observer.* In trivial acts of verbal communication the observer, typically, is the speaker, and this is the one and only reason why PRESENT has been traditionally defined as "the moment of speaking" (Prior 1967; Prior and Fine 1977), reference to which, it has been claimed, is the meaning of the word *now* (Jespersen 1929; Quine 1964; Huddleston 1984; Bache 1985 *inter alia*). But even a cursory glance at the functional properties of *now* shows that this is not so, as *now* can refer not only to present time, but also to past and future time, cf.:

(71) *She now found herself alone for the first time, and the feeling was weird.*

(72) *Now the light was lifting as the sun went down.*

(73) *They will be having dinner now, if I know anything about their habits.*

As has been argued elsewhere (Kravchenko 1990), there is strong evidence that the meaning of *now* is not 'at the moment of speaking' but 'at the moment/time of observation'. Since the role of the observer may not always and exclusively be ascribed to the speaker, *now* can refer to different times of observation, the observers being different individuals. The same is also true about such indexicals as *today, yesterday, tomorrow,* etc. Consider the title of a book by a contemporary Russian writer Boris Vassiljev:

(74) *Zavtra byla voina*
'Tomorrow was war'

The past tense of the verb places the event "war" in the past from the reader's perspective, but the word *tomorrow* does not refer to the reader's future, it refers to someone else's future, and it is understood that this someone belongs to the reader's past. The two indexicals used in the title orient the reader in the two different worlds — the actual world of the reader and the virtual world of the characters whose life stories are told in the book (high school graduates who leave school the day before the war starts). Such construal of virtual worlds in the cognitive domain of linguistic interactions becomes possible as a result of semiotic multiplication of the speaking observer's phenomenological world (see Section 3 of Chapter 1). Consequently, we can speak of a *split temporal perspective* (Kravchenko 1992) in examples like (74): one is the speaker's (reader's), the other is someone else's (another observer's). Schematically, this split temporal perspective is shown in Figure 8 where "Observer$_1$" is the speaker/reader, and "Observer$_2$" is one of the characters in the book. Note that by placing the indexical *tomorrow* at the beginning of the sentence, the author orients the reader as to whose viewpoint is given primacy in the narrative.

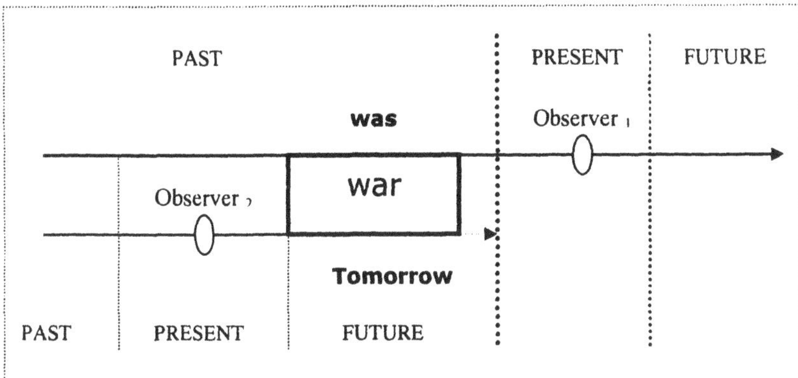

Figure 8. Multiple observers and the split temporal perspective in the linguistic construal of the world

As has been shown in the previous sections, the cognitive distinction between the speaker and the observer is very important and in Russian it is reflected in the grammatical category of aspect. Now, what about aspect in English? Could the overall picture be similar to what we find in Russian?

6.4.2. Aspect Undisguised

Now let us take a look at the English verb. The twelve basic grammatical 'tenses' in the active voice are summed up in Table 10. If we agree that there are three temporal domains in the English world view constituted by categorized perceptual experience (doesn't, after all, the name for each tense begin with either "Past", "Present", or "Future"?), and that grammar reflects some fundamental aspects of human cognition, we must decide what the four groups of finite forms are about. Two approaches are possible: one is structural, and the other conceptual. The structural approach allows to view the four groups of finite forms as an opposition SIMPLE/COMPLEX, whereby simple forms are contrasted with complex forms. Complex Tenses include the Progressive, the Perfect, and the Perfect Progressive, each of which consists of an auxiliary and the participle of the main verb. Since the Simple Tenses are, basically, inflected infinitives (we disregard the Future

Simple here as a historically later development), the opposition SIMPLE/COMPLEX of the finite forms ought to be related to the opposition INFINITIVE/PARTICIPLE of the non-finite forms. It means that, in order to understand the semantic distinction between the simple and complex tenses, we must establish such a distinction between the infinitive on the one hand, and the participle on the other. This is where the conceptual approach to finite forms comes into play.

Table 10. The four groups of tenses in English

TENSE	Simple (Indefinite)	Complex (Definite)		
		PROGRESSIVE be + Ving	PERFECT have + Ven	PERFECT PROGRESSIVE have been + Ving
Present	*I dance*	*I am dancing*	*I have danced*	*I have been dancing*
Past	*I danced*	*I was dancing*	*I had danced*	*I had been dancing*
Future	*I'll dance*	*I'll be dancing*	*I'll have danced*	*I'll have been dancing*

Another name for the Simple Tenses used in grammar books is *Indefinite (Tenses)*. A logically consistent approach to classification would require that the name Definite be used for the Complex Tenses, but then the question arises about the actual meaning of respective terms, *Indefinite* vs. *Definite*. These terms cannot refer to temporal distinctions *per se*, since temporal reference is made explicit through the terms Present, Past, and Future, while reference to concrete time, with different degrees of specification, is achieved by means of different lexical devices as in (75), or as in (76) when the so called 'indefinite' tense is used to refer to very definite time:

(75) *a. He visited Paris in his younger years.*

b. He visited Paris 20 years ago.

c. He visited Paris in 1982.

d. He visited Paris in the May of 1982.

e. He visited Paris during the last week of May, 1982.

(76) *He arrived in Paris at exactly 11 a.m. on the 28th of May, 1982.*

Note, that the INDEFINITE/DEFINITE contrast parallels the SIMPLE/ COMPLEX contrast, so it may be attributed to the contrast between the non-finite forms INFINITIVE/PARTICIPLE as well.

No matter what approach we take looking for the relationships among the English tenses — structural or conceptual — it becomes obvious that an analysis of finite forms and their meaning from the point of view of the underlying cognitive structures should start with the identification of the grammatical status of the opposition of non-finite forms. Surely the fact that virtually every English verb as a lexical item has, apart from the finite forms, two other formal realizations known as the participle and the infinitive, is not incidental. It speaks of some important aspect of structuring knowledge in linguistic forms — the job grammar is called upon to do. Remarkably, the issue of the grammatical meaning of the opposition INFINITIVE/PARTICIPLE, which is an unquestionable instance of grammatical category as metasign, has hardly been addresses in traditional grammar studies.

As has been suggested elsewhere (Kravchenko 1992), the difference between non-finite forms may be described as follows. The infinitive refers to an event (activity, etc.) as an abstract concept, that is, its referent is not something that exists in reality and can be observed as part of a specific situation with its unique spatio-temporal characteristics; in other words, it is not something perceptually *present* in the situation. Rather, it is a mental construct that exists in the unlimited realm of human consciousness, devoid of any spatio-temporal characteristics. It is no coincidence that the term *infinitive* is derived from the corresponding adjective meaning 'having no boundaries or limits, endless or innumerable' (speaking of "dead" metaphors again!).

The grammatical meaning of the participle is reference to an event which exists in reality because it can be observed as part of a specific situation in physical context with its unique spatio-temporal characteristics. The term *participle* is derived from the Lat. *participio* 'take part in sth', that is, its referent is viewed as *being a part of the observed and described situation*. This difference between the infinitive and the participle is reflected in the fact that, whenever we are asked to produce an instance of a verb, we always come up with an infinitive and never a participle — just because a participle would entail a series of questions about *who*, and *when*, and *where* (invoking spatio-temporal characteristics), whereas the infinitive does not. Interestingly, in Russian, for instance, the participle is morphologically marked for gender and number while the infinitive is not, so if a speaker of Russian were to give an example of a verb in the form of the participle, he would be simultaneously telling his listener that the activity is carried out by either one or more than one individuals, and in the case of one individual — that it belongs either to masculine, feminine or neuter gender, thus tying up the referent of the participle to a specific spatio-temporal context. On the part of the listener, were he to hear an isolated participle, the natural unconscious reaction would be to search for the possible referent such as the agent of the activity referred to by the participle, by asking questions like "Who?" and "Where?"

Just like the participle in Russian (and many other languages, for that matter), the English participle has two forms labeled "Present Participle" and "Past Participle". What is the grammatical (cognitive/conceptual) difference between them? As we have seen, *present* means 'that which is before the senses', and *past* is interpreted as 'that which is beyond the senses'. The referent of the Present Participle is, therefore, regarded as something directly observed in the immediate (described) situation, whereas the referent of the Past Participle is regarded as something beyond the immediate (observed and described) situation. In the latter case, the observed and described situation has obvious signs that the event in question took place, but the event itself is not part of the situation, it is not perceptually present to the observer. Thus, when we say something like *The chair is broken*, we attribute a certain feature ("broken") to the object "chair" because we know from experience that structural integrity of this object can change as a result of some event such as "breaking"; so, by using the past participle of the verb *break* we indicate the origin of the feature characterizing the object. More specifically, the past participle implies comparison of two kinds of data accessible at two different times of observation; we compare what we observe now with what we observed before, and it is on the grounds of this comparison that an appropriate inference is made. So, epistemologically, the past participle is a more complex structure for knowledge representation than the present participle. It is not incidental that *–ing* as a grammatical morpheme marking the present participle is among the earliest acquired by English-speaking children, preceding, along with other grammatical morphemes, the past tense/past participle morpheme *–ed* (Brown 1973). This sequence of acquisition (first, the *–ing* forms, then, the *–ed* forms) is an iconic reflection of children's cognitive experience of their immediate linguistic environment. This environment, among other things, is constituted by utterances the caretaker's makes while speaking to the child. Typically, the progressive forms in the speech of parents talking to the child prevail over other forms which contain the *–ed* morpheme, as parents tend to speak about what is taking place at the moment of speech and thus *is observed by the child and themselves*. Strictly speaking, this is a necessary condition for establishing a *cognitive link* between a linguistic form and a corresponding part of the child's environment, when linguistic signs are grounded in the physical context of the communicative act.

The cognitive content of non-finite forms in English is based in the distinction between indefinite and definite sources of information about described events: the infinitive refers to an event or activity without any indication that it is observed by someone, and the participle (either present or past) refers to an event or activity indicating, at the same time, that the source of information about this event may be specified as a particular identifiable observer. As was shown in Section 6.2, the cognitive distinction "indefinite source of information/definite source of information" underlies the grammatical meaning of aspect as a category, which is *expressed by the opposition* INFINITIVE/PARTICIPLE. This distinction *underlies the systemic difference in meaning between the so-called simple tenses and the never-called so 'complex tenses', i. e. the progressive, the perfect, and the perfect progressive.* The formal classification of English finite forms ("simple" vs. "complex") reflects their aspectual distinctions. The simple tenses are, in fact, the Indefinite Aspect Tenses, and they are typically used when we speak about what we know rather than what we observe. The complex tenses are the Definite Aspect Tenses, and they are typically used when we speak about what we observe rather than what we know. Depending on what it is that we actually observe (the process or activity itself unfolding before our eyes, or visible signs that such an activity must have taken place), the Definite Aspect is further specified through special grammatical forms traditionally known as the Progressive and the Perfect.

Situations in which we speak about what is immediately observed may differ considerably, depending on what exactly we do observe (that is, what we can see, hear, smell, etc.). When I say:

(77) *Look, Jack is smoking again,*

I can observe the process itself; Jack is holding a cigarette, periodically putting it in his mouth, inhaling, and then exhaling smoke (frankly, a senseless, hazardous, and very unaesthetic activity). This is a *bona fide* "progressive" situation. Now, if to my question why Jack's face has a greenish pallor I hear an answer such as (78):

(78) *He has smoked three cigarettes in a row,*

I am being informed of the cause of the observed state of Jack's face as a result of the mentioned activity ("smoking"). This is, naturally, a "perfective" situation. In other words, the Perfect is used when we compare what we observe in the situation-of-utterance with what was observed before, and make a certain inference based on this comparison.

Finally, it is often the case that even while we speak about an activity observed in the situation-of-utterance, we, at the same time, compare what we see with what we saw some time before. For example, I see Jack smoking now, and I have also seen him smoking on several occasions throughout the day — in such a case, the two forms expressing different cognitive values are combined yielding the Perfect Progressive:

(79) *Jack has been smoking (all day).*

The Perfect Progressive may be used even though the activity referred to by the verb is not observed in the situation-of-utterance; it is often enough that the observed sign(s) of this activity have a close *temporal* link with it. If Jack's mother, upon his coming home after school, smells a specific odor emanating from his clothes, and asks,

(80) *Have you been smoking again?*

she indicates that she perceives something that is a telltale sign of a very recent activity of smoking, and she compares this with a similar "smoking situation" that took place before (the function of the adverb *again*).

6.4.3. Pedagogical Implications

So, the cognitive content of aspect as a grammatical category of the English verb is related to the contrast between two basic concepts of events (activities, processes, etc.): verbal referents are categorized either as those identified empirically, through direct observation, or those identified on the basis of existing background knowledge. In Chapter 5 these different types of knowledge were characterized as phenomenological knowledge (the observer's competence) and structural knowledge (the speaker's competence). The conceptual contrast between the observer and the speaker is systemic in nature (Kravchenko 1993) and serves as a cognitive basis for many grammatical phenomena we call categories. Recursive use of grammatical categories as metasigns used for the establishment of second-order consensual domain allows indexing and cuing necessary both for the interpretation of observed linguistic behavior in spoken communication and of written texts. When we speak, we speak about *what we see* (hear, feel, etc.) or about *what we just know*. In the first case, the source of information about the verbal referent is definite because the observer may be identified. In the second case no such immediate identification is possible, so the source of information about the event is indefinite. The grammatical meaning of aspect in English is thus defined as "indication to the source of information about the event" which can be definite (based on observation), or indefinite (based on the speaker's knowledge). In this, aspect in English does not differ noticeably from aspect in Russian as analyzed in the previous sections, either conceptually or structurally, because in both languages the grammar of aspect is rooted in similar conceptualizations of events which are categorized relative to the *identifiable observer as a source of information.* Both languages have a tripartite system of verb forms; one type of form expresses the indefinite aspect (IMP_2 verbs in Russian and the simple tenses in English), while two other types express the definite aspect (IMP_I and perfective verbs in Russian, and the progressive and perfect tenses in English). Such conceptual parallelism helps simplify the process of tense-aspect acquisition by Russian learners of English, when a simple three-step procedure may be used for choosing a correct form in any discourse situation, as shown in Figure 9.

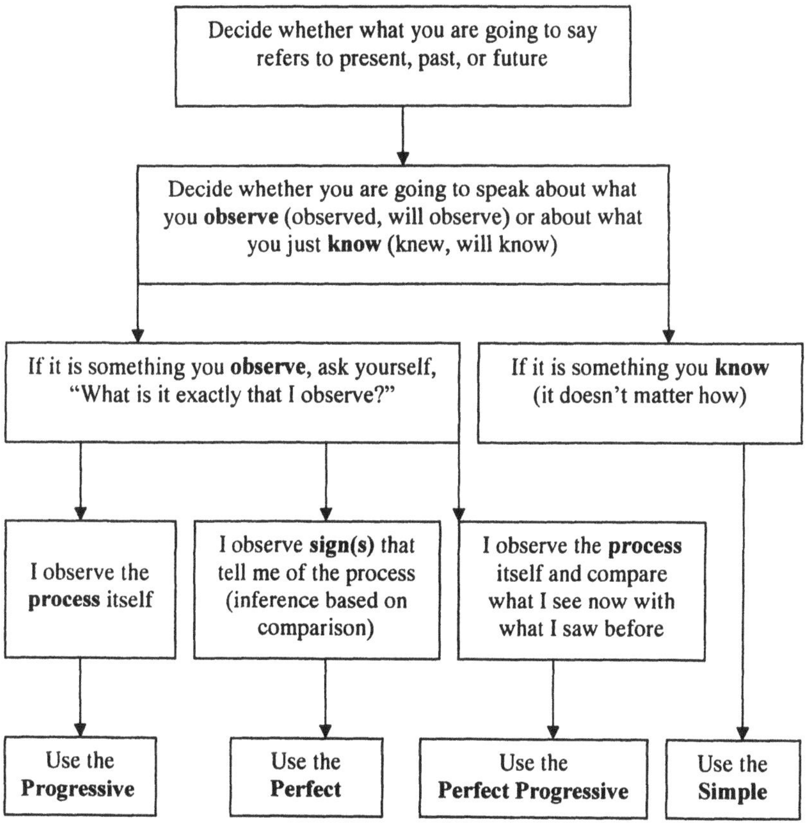

Figure 9. Decision tree for choosing a tense

<u>Step One</u>. The speaker must decide whether what he is going to say refers to his present, past, or future, thereby establishing the frame of temporal reference. It is important to remember here that in making such a decision the speaker *should*

not be guided by the grammatical form of the verb he would use in his native language, as it might be misleading. Instead, he should make recourse to the cognitive concepts of PRESENT, PAST, and FUTURE as they have been specified above.

Step Two. The speaker needs to answer the question, "Is what I am going to say based on somebody's observation, or is the source of my information unspecified?" In the latter case, a positive answer results in the choice of the Indefinite Aspect Tense (that is, one of the simple tenses), and that is all there is to it. In the former case, a positive answer leads to a choice of one of the Definite Aspect Tenses (a progressive or a perfect tense), and so one more step is necessary.

Step Three. The question the speaker needs to ask is, "What is it that I (he, them, etc.) observe, exactly? Is it the process itself unfolding before the observer's eyes, or is it only some sign(s) telling the observer that this process must have taken place?" Correspondingly, in the first case one would use the progressive verb, and in the other case, the perfect.

But what about the Perfect Progressive? Nothing much, really. It's Perfect and Progressive combined, both conceptually and structurally: in a prototypical situation, the Perfect Progressive will be used when conditions for the use of Perfect and Progressive obtain at the same time. Thus, talking to a friend at the library, I may point at someone we both know and say:

(81) *Just look at him! He's been sitting at that table for a good 5 hours!*

I use the Present Perfect Progressive here because:

(a) the situation I describe is part of my present ("before the senses"),
(b) my statement is based on what I observe ("*Look at him!*"),
(c) the process I describe is part of the observed situation ("participates in it"), and
(d) I compare what I observe ("have") now with what I observed ("had") some time before.

This simple procedure, once its conceptual justification has been explained to and understood by Russian-speaking EFL students, allows to dramatically improve the process of grammar acquisition, ridding it of semantically vague interpretations and different kinds of "exceptions" based on them. Moreover, the cognitive difference between the infinitive and the present participle, once explained to, and understood by, the student, provides a sound and comprehensible explanation of the difference in meaning and function in the case of constructions with clausal arguments (*I saw him dance* vs. *I saw him dancing*), relating the difference between the two structures to different types of knowledge categorized in linguistic form (see Chapter 7).

The cognitive approach to tense and aspect in English outlined above has been successfully used by the author and some of his colleagues and graduate students in teaching EFL to speakers of Russian at various educational institutions for over 10 years now (Kravchenko 1997). As accumulated experience shows, it takes about one to two months for the average student to understand the "works" of the entire tense-aspect system, that is, the meaning and function of all the tenses — the rest is a matter of practice.

6.4. Summary

Orthodox theoretical grammar has been unable to account for various intriguing phenomena related to Russian aspect both on the morphological and syntactic levels. Moreover, it has not even concerned itself with probing into the subtle cognitive mechanisms which underlie conceptualization processes reflected in grammatical categorization. As I have argued, failure of orthodox grammar to facilitate advances in applied linguistic tasks calls for alternative accounts of grammatical phenomena that seem to have been extensively explored, analyzed and described, as the outcome of such linguistic endeavors may not be described as satisfactory. No wonder that the 'terrain' of tense-aspect categories remains a large 'white spot' on the cognitive map of natural language grammar..

Counter to orthodox views on the meaning of aspect, real world situations do not possess 'internal temporal structure' for the simple reason that 'temporality' is a subjective notion grounded in humans' perceptual experience which tends to be holistic rather than analytic. As has been argued in this chapter, linguistic categorization of a situation by means of an aspectually (un)marked verb does not depend on an analysis of the situation structure, but on conceptualizing a situation as perceptually 'present' or 'non-present'. A cognitive approach to verbal aspect, by taking into account the experiential origin of meaning, assumes it to be the grammatical categorization of a *primary cognitive structure related to perceptual experience* (Stoevsky 2001; Kravchenko 2002b). In Russian, this primacy of aspect governs the morphological paradigm for verbal tense by constraining the so-called perfective verbs for present tense and is reflected in the cognitive/functional asymmetry of participles. Referential asymmetry consistently displayed by participles in certain contexts, and the 'other's point-of-view effect' created by selective use of either present or past participle, pose a problem for functional-semantic accounts in traditional theoretical grammar. Using a cognitive analysis of the meaning and function of the present and past tense active participles in the attributive function, I have shown that the traditional treatment of these forms in Russian grammar (in terms of temporal reference attributed to finite verbs) should be assessed as inadequate. By taking a consistently cognitive stance ("everything said is said by an observer to another observer"), I have shown that the grammatical status and function of Russian active participles cannot be fully understood and/or appreciated without relating them to the function of knowledge representation as the primary function of grammatical categories. In English, perceptual groundedness of grammatical categories accounts for the specific uses of different types of verbs in the simple versus complex tenses as a function of viewpoint (Głaz 2002; 2007) in knowledge representation (Brisard 2005).

The suggested cognitive framework for Russian aspect parallels a similar framework for English (Kravchenko 1990; 1992; 1997; 2002b), providing a unified methodology for cognitively oriented typological tense-aspect studies in different languages. By applying a similar procedure in the analysis of grammatical meaning of the so-called English tenses, I have shown that their grammatical function is very much the same as the function of aspectual contrasts in the morphology of the Russian verb. Just like in Russian, aspect in English is an indexical device, a metasign, used to orient communicating individuals in regard to the source of knowledge categorized in linguistic form. As such, grammatical aspect reflects a universal cognitive mechanism which governs languaging as recursive behavior in a consensual domain of interactions, and — not surprisingly — it is structured similarly both in Russian and in English. Undoubtedly, this is one of the major reasons why native speakers of Russian may effectively acquire English grammar. A cognitive account of grammatical categories, their meaning and function is an effective alternative to the traditional semantic account, giving deeper insights into the cognitive nature of grammar and enhancing the process of its acquisition.

CHAPTER 7

INTERPRETING LINGUISTIC STRUCTURE: HOW WE KNOW WHO DOES WHAT WHEN

7. 1. Complex Sentence as a Cognitive Structure

7.1.1. Introduction

While the traditional approach to grammar attempts to provide an explicit set of features of meaning 'expressed' by a given form, and the rules for combining and recombining such forms to express meanings as 'truths' about the 'real world', in contemporary cognitive research these features are best viewed as assemblies of concepts as cognitive structures, which are construed with the help of language and 'acquired' via natural language. The outcome of this 'acquisition process' is the sum total of our linguistic experiences; viewed as a system of categorized patterned cognitive experience, grammar includes what traditionally has been referred to as linguistic and extralinguistic phenomena — such as perceptual, for example (Johnson-Laird 1983). This categorized patterned cognitive experience as components of a consensual domain applied recursively without the consensual domain is, basically, what we call *knowledge*. Note that in this sense knowledge presupposes (linguistic) *interactions*; it isn't some objective truth about the 'real world' waiting 'out there' to be discovered, it is *live action* described by a languaging observer as a function of fine structural coupling of an organism with the environment. To understand the grammar of a particular language is to understand what cognitive mechanisms for categorizing experience are at play, and what form such categorization takes in natural language on the one hand, and in written language, on the other.

Because grammar is a sign system for storing categorized cognitive experience (knowledge) wherein every grammatical category relates to some essential aspect of cognitive processing, the principal objective of grammatical theory should be explication of the relationship between linguistic structures (grammatical constructions) and cognitive concepts behind them (Taylor and MacLaury 1995; Shibatani and Thompson 1996; Heine 1997). A basic set of such concepts must not be very large or radically different from language to language, as phenomenology of cognition is largely independent of racial or ethnic factors. In other words, grammar is called upon to provide a means for representing *cognitive experience* via conceptualization and categorization (Svorou 1994). On the level of syntactic structure such representation is associated with *iconicity* (Haiman 1985; Givón 1995), when syntactic properties of sentences are analyzed from the point of view of *information flow*.[24] However, the term *information flow* may refer to two cognitively different processes. On the one hand, the linear structure of the sentence or its parts may reflect the order in which the described situation is processed *perceptually*, with perceptually salient elements becoming *figures* as opposed to *ground* (Tomlin 1987). A trivial example would be the particular order of adjectives used in strings of attributes preceding a noun in a noun phrase, as in (82):

(82) a. *Suddenly I was approached by a tall young man.*

b. ?? *Suddenly I was approached by a young tall man.*

Such perceptual salience may be called *salience of the first order*. On the other hand, the linear structure of the sentence may depend on *salience of the second order*, when the speaker assigns to a thing or property a degree of significance determined by the speaker's relative system of values rather than by perceptual salience, as in (83):

(83) A: — *What's your new neighbor like?*

B: — *Well, John Bradley is a rather young, tall man...*

Examples like (82) and (83) illustrate an important distinction between two perceptual/cognitive types of discourse: we speak either about *what we perceive* while we engage in first-order activity, or about *what we know* as a result of our experience integrated into our system of beliefs and values represented, at the neural level, by networks of concepts (see Chapter 3).

In mainstream cognitive science concerned with the issue of language 'processing' one of the most important tasks is to determine the source to which a given concept, either at the level of oral discourse or written communication, can be traced (cf. Hill 1994). Because "everything said is said by an observer", the ultimate point-of-reference for any concept is *sensory data input*. However, while languaging, that is, when acting in a second-order consensual domain, human individuals try to orient others with regard to their first-order consensual domains, and they do so with the help of metasigns, or grammatical categories, which cue understanding by indexing relevant aspects of both first- and second-order consensual domains. While in spoken communication such cuing relies on various characteristics and parameters of the physical context, including the complex cog-

[24] Although in the previous chapters such concepts as "information", "meaning", "expression of thought" etc., widely used in traditional linguistic analysis, were criticized as misleading, in the context of this discussion I find it possible to use these terms as they have been used in linguistic and philosophic literature — simply to avoid additional difficulties in making the main point clear.

nitive dynamics of linguistic behavior, in writing it is achieved with the help of symbolized linguistic signs themselves. So, for the reader to experience any orientational effect from reading a piece of writing, he must reconstruct the first-order consensual domain indexed by written symbols of linguistic signs organized into larger units such as sentences and texts.

The first part of this chapter addresses the problem of relationship between the form and meaning of the complex sentence of the type *When S, S* in the context of admissible structural variations — involving both morphological and syntactic features — and their relevance for the interpretation of sentence meaning.

7.1.2. The Data

Certain theoretically possible sequences of written words, described in grammars as sentences of a particular structural type, have been analyzed. These sentences were grouped in three sets depending on the tenses used in the clauses: in Set I the verbs in both clauses are in the past simple tense, in Set II there is a variation of the past simple and past progressive, and in Set III it is the past simple and the past perfect. Note that in all the sentences the verbs are in the past tense; it is with regard to expression of aspectual meaning that their forms vary, from the Indefinite Aspect in the simple forms to the Definite Aspect in the complex forms (progressive and perfect). For convenience, to help see certain tendencies in similarly structured sentences, all examples from Sets I through 3, apart from being numbered, have been supplied with an alternative notation.

Set I.

(84)	a.	? *When John came home, Ann cooked.*	Ia_1
	b.	*When John came home, Ann cooked dinner.*	Ia_2
(85)	a.	? *Ann cooked when John came home.*	Ib_1
	b.	*Ann cooked dinner when John came home.*	Ib_2
(86)	a.	* *John came home when Ann cooked.*	Ic_1
	b.	? *John came home when Ann cooked dinner.*	Ic_2
(87)	a.	?? *When Ann cooked, John came home.*	Id_1
	b.	?? *When Ann cooked dinner, John came home.*	Id_2

Set II.

(88) a. *When John came home, Ann was cooking.* IIa_1

b. *When John came home, Ann was cooking dinner.* IIa_2

(89) a. *Ann was cooking when John came home.* IIb_1

b. *Ann was cooking dinner when John came home.* IIb_2

(90) a. ? *John came home when Ann was cooking.* IIc_1

b. *John came home when Ann was cooking dinner.* IIc_2

(91) a. ? *When Ann was cooking, John came home.* IId_1

b. *When Ann was cooking dinner, John came home.* IId_2

Set III.

(92) a. ?? *When John came home, Ann had cooked.* $IIIa_1$

b. *When John came home, Ann had cooked dinner.* $IIIa_2$

(93) a. ?? *Ann had cooked when John came home.* $IIIb_1$

b. *Ann had cooked dinner when John came home.* $IIIb_2$

(94) a. ?? *John came home when Ann had cooked.* $IIIc_1$

b. ? *John came home when Ann had cooked dinner.* $IIIc_2$

(95) a. ?? *When Ann had cooked, John came home.* $IIId_1$

b. ? *When Ann had cooked dinner, John came home.* $IIId_2$

These sample data were obtained with the help of a questionnaire offered to a number of educated adult native speakers of English who were asked to assess the sentences as either "normal", "doubtful (?)", "highly doubtful (??)", or "unacceptable (*)". Although there was a slight dispersion of answers for the latter three categories, the tendency to mark down certain types of sequences as deviant appeared to be quite obvious. The data are summed up in Table 11.

Out of the 24 sentences in the sample, fourteen, or almost 60%, have been marked down as doubtful. Out of the ten sentences marked as normal four have the structure *When* S_1, S_2 and four — S_2 *when* S_1. The main difference between the syntactically similar S_1 and S_2 in our sample is that while in S_1 the verb is in the simple form in all the examples, in S_2 it varies from simple to progressive to perfect. Note that most of the doubtful sequences in Table 11 (more than 70%) are the sequences where S_2 is pre-posed by *when*. Obviously, there are some constraints on the combinability of *when* with a clause, which seem to be imposed by the kind of aspect expressed by the grammatical form of the verb used in the clause. And of course, the question is, "What is going on here?"

Table 11. The degree of acceptability of possible sentence structures

Degree of Acceptability	Sentence Structure			
	When S_1, S_2	S_2 *when* S_1	*When* S_2, S_1	S_1 *when* S_2
Normal	Ia_2 IIa_1 IIa_2 $IIIa_2$	Ib_2 IIb_1 IIb_2 $IIIb_2$	IId_2	IIc_2
Doubtful	Ia_1	Ib_1	IId_1 $IIId_2$	Ic_2 IIc_1 $IIIc_2$
Highly Doubtful	$IIIa_1$	$IIIb_1$	Id_1 Id_2 $III\ d_1$	$IIIc_1$
Unacceptable				Ic_1

As the data in (84) – (95) show, the constraints seem to depend on an interplay of the following factors:

(i) the actual order in which the clauses are combined (S_1 S_2 or S_2 S_1),

(ii) the relative position of *when* (*when* S_1 vs. *when* S_2),

(iii) the kind of aspect (Indefinite or Definite) expressed by the grammatical form of the verb used in each clause,

(iv) presence/absence of a verb complement in a clause.

So let us take a closer look at these factors.

Clause Combining and the Position of When. In view of the critique of the traditional definition of sentence given in Chapter 5, and to disambiguate the meaning of the term *sentence* as compared to the term *utterance*, I will use the former to refer to a sequence of written words which is regarded as capable of standing alone to express a coherent thought. In our case it means that the syntactic structure of the type *When S, S* should express a coherent thought to qualify as a sentence. However, as our sample shows, to educated native speakers not all possible sequences with such structure possess the necessary degree of coherence. For some reason, certain sequences were felt by native speakers to be less acceptable than others.

If we look at the (propositional) content of the two clauses S_1 and S_2 (*John came home* and *Ann cooked dinner*), there doesn't seem to be anything there to tie one to the other in terms of inherent semantic properties accounting for the coherence of thought expressed by the complex sentence as a grammatical unit. A sequence of two simple sentences such as in (96) would iconically reflect the sequence of respective events as they took place from the point of view of an observer who happened to be there at the time and now, as a speaker, is sharing his knowledge with others:

(96) *John came home. Ann cooked dinner.*

Whether Ann cooked dinner *because* John came home, or it was just a coincidence, remains an open question as long as there is no indication to contextual parameters or background knowledge necessary for the adequate interpretation of the relationship between the two events. Similarly, a sequence such as in (97) could be interpreted either causally, as 'John came home because Ann cooked dinner', or simply iconically as two unrelated events occurring successively:

(97) *Ann cooked dinner. John came home.*

To show that a certain relationship really exists, conjunctions are used to specify this relation either as subordinate or coordinate (or hypotactic and paratactic — we are not concerned with the terminological issue here). Thus, with an addition of *when,* the clause *John came home* becomes subordinate to *Ann cooked dinner* in (84b), and vice versa in (85b). This relation of subordination means that in the first case John's coming home is perceived as less relevant than Ann's cooking dinner, while in the second case it is the other way around. But what, then, is the difference between (84b) and (85b) in terms of their informative value? If both possess the same degree of iconicity reflecting the perceptual processing of information flow, how can the different clausal order be explained? This is the first question I will attempt to answer by discussing the relevant data in Table 11.[25]

[25] It must be noted here that the explanation of the difference in clause sequencing in terms of manifestation of their *thematic status* (Halliday 1967; Fries 1983; Matthiessen and Thompson 1988 *inter alia*) does not answer the question how a particular thematic status is assigned in the first place, so I am not going to probe into that area.

Verbal Aspect and Complementation. Now, what kind of relationship exists between verbal aspect and complementation in, say, S_2 on the one hand, and the resulting acceptability of the complex sentence in relation to clause combining, on the other? Upon a closer look, a certain pattern emerges. If the verb has no complement in S_2, only two normal complex sentences (that is, sentences expressing a coherent thought) are possible with the structure *When* S_1, S_2 and S_2 *when* S_1 — (88a) and (89a). All the four interplaying factors considered, this can be put down only to the form of the verb (the progressive) and the aspectual meaning expressed by it. As soon as the progressive is replaced by the simple or perfect, the resulting complex sentence begins to lack coherence necessary to view it as an acceptable structure for representing knowledge. However, if the verb is followed by a complement (*dinner* in our examples), the constraint on sentence acceptability for the structures *When* S_1, S_2 and S_2 *when* S_1 disappears.

Moreover, the progressive is the only form which, when used in S_2 with a complement, yields acceptable complex sentences with the structures S_1 *when* S_2 and *When* S_2, S_1. Both the sequences *When* S_1, S_2 and S_2 *when* S_1 are acceptable regardless of the kind of aspect (Definite or Indefinite) used in S_2 as long as the verb takes a complement. However, the sequences *When* S_2, S_1 and S_1 *when* S_2 yield only one acceptable sentence each, again with the progressive verb in S_2. So, the second question I will attempt to answer is, "What cognitive mechanisms are at work here?"

7.1.3. The Hypothesis and Discussion

The data presented above allow to formulate the following hypothesis:

> *Constraints on certain sequences in the complex sentence with an adverbial* when- *clause are the result of cognitive discordance between the indexical meaning of verbal aspect and the conjunction* when *on the one hand, and clause sequencing which iconically reflects the cognitive processing of perceptual data, on the other.*

In keeping with the basic principles of the theory of linguistic indication (Kravchenko 1992), indexical phenomena in language, both on the morphological and syntactic levels, cue the processing of linguistic structures as signs and metasigns which represent categorized (perceptual) experience in the first-order and second-order consensual domains. The primary point-of-reference for indexical phenomena in oral discourse is the languaging observer who categorizes physical, biological, social, and cultural aspects of the environment relevant for the interpretation of his linguistic behavior by others, and he does so by using linguistic signs grounded in his first-order consensual domain (domain of physical interactions), and his second-order consensual domain (relational domain of linguistic interactions). Importantly, in the case of written texts there is no first-order consensual domain, so the writer symbolizes, and the reader interprets symbolizations of, physical signs (words) by drawing on their individual experience of natural linguistic signs. Basically, it means that the words symbolized by the writer in written form may be grounded differently from the words the reader takes written symbolizations to be encodings of — it all depends on the writer's and reader's individual developmental histories. However, because of the second-order consensual domain in which the writer and the reader interact as members of a community speaking a particular language, grammatical categories, or metasigns they use are, to them, grounded in the relational domain of linguistic interactions which, as was argued in Chapter 5, sustains the unity of a community as a third-order living system. Thus metasigns supersede lower level signs as categorization devices responsible for cuing 'general' understanding of who did what when and how it became known to the speaker — just like a third-order living system (a community of individual human organisms) supersedes a second-order living system (an individual human organism). The actual source of knowledge 'represented' in linguistic structures may not always be the speaker; although the speaker is always the observer at the same time, the reverse is not necessarily true, particularly in written texts as illustrated by Figure 8 in Chapter 6.

Every time an indexical is used in discourse, whether oral or written, it points to an identifiable source of information whose perceptual experience is integrated as a semantic component in the meaning of the utterance or sentence as a whole. Examples from our sample contain three types of indexicals: the tense, the aspect, and the conjunction *when.* Since there is no tense variation affecting the interpretation of the sentence, this indexical is not considered here. As for the morphosyntactic category of aspect, it was shown in Chapter 6 that its cognitive content is related to the contrast between two basic concepts of events, and the grammatical meaning of aspect consists in indicating the source of information about the event which can be definite (based on observation), or indefinite (based on the speaker's knowledge, the source of this knowledge left unspecified).

The lexical item *when* in traditional functionally oriented grammars (Quirk et al. 1973; Leech and Svartvik 1975; Givón 1993 *inter alia*) is assigned different roles and defined either as an adverb (*When will he leave?*), subordinating conjunction (see our sample), or relative pronoun (*Since when has this been going on?*). However, functional-syntactic differences aside, one cannot rid oneself of the feeling that in all three cases there is some basic sameness in meaning in the sense that there is one core concept (cognitive structure) underlying the meaning of *when* regardless of its syntactic function. Following the tenet of the biocognitive philosophy of language about the experiential nature of meaning, and in view of the fact that *when* is traced etymologically to the Gothic accusative of *who* (Skeat 1956) and thus belongs to the pronominal class (interrogative pronoun), the meaning of *when*, as a conceptual structure, may be understood only in the context of its *indexical properties,* that is, *its role in identifying the source of information/knowledge* categorized in the complex sentence of which it is a part.

By pre-posing *when* to a clause, the speaker indicates that at least part of the information contained in the sentence has an identifiable source. But who is this source in our examples, John, Ann, or the speaker/writer? As a matter of fact, the question "Whose point of view is reflected in the sentence?" was also in the questionnaire used to obtain the data; interestingly, almost all the respondents said it was the speaker's in all the sentences. However, were it really so, and were the speaker (or the so-called 'moment of speech') the point-of-reference of indexical temporal expressions as claimed in traditional approaches to grammar (Jespersen 1929; Hudleston 1984; Bache 1985), why would the very same respondents mark certain sentences as doubtful? Obviously, the speaker is not always at liberty to choose any of the possible linguistic structures for 'expressing' a particular thought in a coherent way. This coherence depends on the interplay of the four factors mentioned in Section 2. Specifically, there seems to be a certain dependency of sentence acceptability on the relative position of *when* with respect to aspect expressed by the verb in the clause. Let us compare structurally similar sentences from the three sample groups.

In Set II, ($88a/IIa_1$), ($88b/IIa_2$), ($89a/IIb_1$), ($89b/IIb_2$), ($90b/IIc_2$) and ($91b/IId_2$) have been all checked as normal as if it really did not matter which clause with the verb in what aspect in what sequence was pre-posed by the conjunction *when.* But the picture changes as soon as the progressive form of the verb in S_2 is changed either to the simple as in Set I, or the perfect as in Set III. Sentences ($88a/IIa_1$) and ($89a/IIb_1$) from Set II become doubtful in Set I and highly doubtful in Set III; sentence ($90b/IIc_2$) becomes doubtful as ($86b/Ic_2$) and ($94b/IIIc_2$), and sentence ($91b/IId_2$) becomes doubtful as ($95b/IIId_2$) and highly doubtful as ($87b/Id_2$). Moreover, sentences marked as doubtful change their degree of acceptability: ($91a/IId_1$) from doubtful becomes highly doubtful both as ($95a/IIId_1$) and ($87a/Id_1$), and ($90a/IIc_1$) from doubtful becomes highly doubtful as ($94a/IIIc_1$) and unacceptable as ($86a/Ic_1$). How can this phenomenon be explained?

Let us begin on a hunch that if the two propositions expressed by S_1 and S_2 are combined in one sentence of the type *When* S_1, S_2 to form a 'coherent thought', there must be something that allows to do so. According to the hypothesis, and taking into account sentence iconicity, it is indication that the information contained in the main clause is the result of the processing of perceptual data by an experiencer identified in the subordinate clause. Literally, the meaning of ($88a/IIa_1$) may be described as follows: 'When John came home, this is what he saw' (i. e. 'he saw Ann cooking'). The progressive verb in S_2 indicates that the event 'Ann's cooking' is described as observed by someone, and *when* indicates that the observer is John. This iconic property of the sequence *When S_1, S_2* accounts, probably, for the fact that the alternative sequence *S_2 when S_1* is believed (at least, by the testees who took part in the experiment) to be less frequent. However, it is not so infrequent in written discourse, where it has, presumably, a specific grounding function. By placing the main clause with the progressive verb before the *when*-clause, as in (89a/IIb_1) and (89b/IIb_2), the speaker/writer indicates that the source of information about the event 'Ann's cooking' is Ann herself, and the situation described in the first clause is viewed as if from Ann's perspective or vantage point (MacLaury 1995; 2002), while in the second clause the observer is still John. This results in a double, or split, cognitive perspective on the complex situation as a whole discussed in Chapter 6 and falls within the scope of stylistic phenomena of *personification*.

Now, why in Set II both (88a/IIa_1) and (88b/IIa_2) have been marked as normal, and in Sets I and III absence of the complement in S_2 results in anomalousness? The difference between (88a/IIa_1) and (88b/IIa_2) is only in what may be called the 'degree of informativeness' of S_2: in the former, John sees enough to identify Ann's activity as cooking, but he cannot tell what it is that Ann is cooking; in the latter, he evidently sees enough to infer that it is dinner that is being cooked. The perfect in (92a/$IIIa_1$) and (92b/$IIIa_2$) indicates that John sees something and compares it with a previous state of affairs: seeing dinner ready and waiting, he naturally infers that it has been cooked by Ann (example (92b/$IIIa_2$)), while in (92a/$IIIa_1$) it is indicated that John sees something as a result of Ann's cooking but what it is that he sees is left unspecified, creating a semantic lacuna.

A similar semantic gap is found in (84a/Ia_1). If we compare (84b/Ia_2) and (88b/IIa_2), we will see that, because of the different aspects used in S_2 in these examples, the informative value of the sentence differs respectively: *Ann was cooking dinner* represents phenomenological knowledge, it describes the situation as 'here-and-now', that is, at the time of John's observation, so there is no causal relationship between the two propositions. By contrast, *Ann cooked dinner* represents structural knowledge, it describes the situation as a structural part of the world known to the speaker as a result of *accumulated experience* of similar situations; a functional structure is characterized by *causal* relationships of its components, so in (84b/Ia_2) the proposition in S_1 stands in a causal relationship to the proposition in S_2 yielding the following meaning: 'Ann cooked dinner *every time* John came home so he could eat'. Omission of the complement in (84a/Ia_1) creates a slight semantic anomaly *because* of the causal relationship between the two parts, as in the absence of the complement it is not quite clear *why* Ann should start cooking at all every time John came home.

Sentences (90a/IIc_1) and (91a/IId_1) are perceived as anomalous for the simple reason that by pre-posing S_2 with *when* the speaker indicates that Ann is the observer of the event 'Ann's cooking', yet absence of the complement creates a similar semantic lacuna which is also obvious in (94a $IIIc_1$) and (95a $IIId_1$) when compared with (94b $IIIc_2$) and (95b $IIId_2$) as to the degree of acceptability. At the same time, all the (c) and (d) sequences in Sets I and III have been marked as anomalous. According to our hypothesis, in sequence (d) *when* identifies Ann as the observer of the event 'John's coming home', while the indefinite aspect of the verb used in S_1 indicates that the information about this event is part of the speaker's background knowledge obtained in an unspecified manner (the speaker knows for a fact that John came home, but the source of this knowledge is unspecified). Thus, the specific relative positioning of the two indexicals in the complex sentence of the type *When* S_2, S_1 results in *cognitive discordance* with sentence iconicity inasmuch as identification of the perceptual processor of information flow is involved. Consequently, there is no sufficient *cognitive link* between the two clauses that would allow to view the whole sentence as a linguistic structure representing a quantum of knowledge (a complete thought) in a coherent way.

As I have tried to show, the cognitive value of different acceptable sequences of clauses in a complex sentence of the type *When S, S* may be assessed from the point of view of the relationship between their indexical and iconic properties. The suggested hypothesis may provide a useful tool for the analysis of sentential and, by extension, written discourse grammar by making recourse to the concept of *cognitive processor* of the information flow iconically represented by the syntactic structure of the sentence. I have argued that the order in which clauses in a complex sentence are combined is influenced, on the one hand, by an interplay of such cognitive factors as the speaker and the observer identified as possible sources of information (knowledge) about the described events and, on the other hand, by the indexical meaning of the pronominal *when* and the markedness of definite aspect verbs for event observability. So it may be concluded that, in realizing their cuing function in the interpretation of either discourse or written text, grammatical categories as metasigns of different kinds (such as iconic and indexical in our examples), or as orientational devices, must not diverge with regard to the observer as the point-of-reference in a particular situation categorized in the form of a sentence or utterance. If they do, the resulting effect is one of semantic anomalousness, accounting for incoherence of the thought thus 'expressed'.

The discussion of specific data about one particular type of complex sentence offered above is by no means exhaustive and does not allow at this point to make broader generalizations encompassing other types of sentences. However, I believe that such generalizations are, in fact, quite possible — but of course, only further research into the matter may show if my assumptions are valid. In the meantime, I suggest that we take a look at another kind of syntactic structure in order to see if what has been claimed so far about grammar as a means to structure linguistically categorized experience applies to a very interesting type of constructions with infinitive/participle clausal arguments.

7.2. The Cognitive Distinction in Infinitive/Participle Clausal Arguments

7.2.1. The Problem

Those learning or teaching English as a foreign language have always been baffled by the hard-to-grasp differences in meaning between the infinitive and the participle in certain types of constructions, as in (99) – (100):

(98) a. *John heard someone whistle in the dark.*

b. *John heard someone whistling in the dark.*

(99) a. *I saw him cross the street.*

b. *I saw him crossing the street.*

(100) a. *We noticed them leave.*

b. *We noticed them leaving.*

Some grammars attempt to explain the difference between such constructions by making recourse to the notion of aspect understood in orthodox grammar as the morphologically sustained contrast between the progressive and non-progressive aspect, or between the ongoing/repetitive and one-time actions (Quirk et al 1973). Indeed, such difference may be observed in certain contexts where adverbials serve exactly this purpose — to help see the difference between repetitive and one-time actions, as in (101):

(101) a. *I heard the door slamming all night long.*

b. *I heard the door slam just after midnight.*

A similar stance is taken by Martin Hewings (1999: 78):

(102) a. *Did you hear those dogs barking most of the night?* (the action is repeated or happens over a period of time)

b. *I noticed him throw a sweet wrapper on the floor, so I asked him to pick it up.* (the action happens only once)

(103) a. *I was able to watch them building the new car park from my office window.* (we watch etc. some of the action but not from start to finish)

b. *I watched him climb through the window, and then I called the police.* (we watch, hear etc. the whole action from its start to its finish)

However, in certain contexts the difference in meaning between the infinitive and participle constructions may not be so obvious:

(104) a. *As I listened, I heard a door slamming and a strange noise coming from the lower deck.*

b. *As I listened, I heard a door slam and a strange noise come from the lower deck.*

Moreover, as soon as the sentence is reduced to a minimal syntactically acceptable structure, the issue ceases being unambiguous altogether, as in (105):

(105) a. *I heard the door slamming.*

b. *I heard the door slam.*

Examples similar to (102 5) – (103 6), usually found in grammar books, lean heavily on the linguistic context in explicating the difference between the two competing constructions. However, should we delete the modifier and change the determiner in the noun phrase in (102a 5a), we would come up with two equally possible alternatives:

(106) a. *Did you hear the dogs bark?*

b. *Did you hear the dogs barking?*

As for (103), in a changed context — as in (107) – (108) — the difference between the infinitive and the participle is also not so obvious:

(107) a. *I watched them build the new car park across from the office.*

b. *I watched them building the new car park across from the office.*

(108) a. *I watched him climb up the steep precipice.*

b. *I watched him climbing up the steep precipice.*

Infinitive and participle constructions with verbs of sense perception may be used without any accompanying contextual information that could help see the difference in meaning, as in examples (109) – (111):

(109) a. *I saw her dance.*

b. *I saw her dancing.*

(110) a. *I heard him speak about that.*

b. *I heard him speaking about that.*

(111) a. *He watched me move down the narrow ledge.*

b. *He watched me moving down the narrow ledge.*

This allows some authors to claim that there isn't any important difference between the two phrases in such cases (Murphy 1985). So, the question one is tempted to ask is, "What guides the speaker in making the choice in favor of one construction rather than the other?" Surely this choice cannot be random or cognitively irrelevant, and that grammars cannot provide a coherent explanation does not mean that there is no relevant difference between the two types of construction. Most probably, it means that the approach used, which I call 'traditional semantic', is not the one suited for the task.

When grammarians focus on explaining away the difference between the infinitive and participle constructions in question, the usual interpretation is as follows. The infinitive expresses the idea that the activity was witnessed by an observer/speaker from beginning to end, while the participle indicates that the observed subject, e. g. *she* in (109b), was in the middle stage of the performed activity and the observer did not see it through to the end (Palmer 1968; Vendler 1968). Should this approach be understood in the sense that the infinitive somehow possesses an ability to express the idea of completion of an activity or process, while the participle possesses an ability to express its incompletion? It is just such understanding that makes orthodox linguists relate the difference in meaning between these two types of construction to the notions of progressive or non-progressive aspect. But in such a case the first thing to be done is to clearly define the problem and the methodology used for its solution. So, what is the problem?

In a nutshell, it is the grammatical status of the non-finite verbs which go under the names *infinitive* and *participle*. On the one hand, if we approach these forms systematically, they constitute a classical morphological opposition since there is not a single verb in English, except the modals, that wouldn't have both the forms. On the other hand, although in descriptive linguistics a regular morphological opposition of two forms of a word has always been a telltale sign of a specific grammatical meaning expressed by it, and thus of a corresponding grammatical category, contemporary English studies have shunned the question, "What is the nature and meaning of the grammatical category expressed by the opposition INFINITIVE/PARTICIPLE?"

Still, whatever the answer to this question, it would not be enough to resolve the problem of semantic equivalence of clausal arguments in infinitive/participle constructions. Typically, when these constructions come in the focus of discussion in grammars, the whole issue boils down to the semantic difference between the infinitive and the participle, becoming a classical example of reductionism. But because a construction itself carries meaning, we should abstain from reductionism and analyze the construction in question as a whole. Viewed holistically, it would have the following structure, with the second noun phrase being optional and thus irrelevant for our analysis: VP + NP + Infinitive/Participle [NP]. A consistent analysis of the meaning of this construction must include the relationship between the main finite verb, the noun phrase, and the non-finite verb, revealing an interplay between the construction elements at a cognitive level. In what follows, I will analyze such interplay between the elements of the construction VP + NP + Infinitive/Participle in order to show how the construction in question categorizes different types of knowledge.

7.2.2. What Non-Finite Clausal Arguments Describe

Let us look at some of the Section 7.2.1 examples from the point of view of the relationship between the linguistic form of the construction and its corresponding cognitive structure. By *cognitive structure* I understand a structure for categorizing and organizing knowledge about the situation described in the sentence. Since grammatical phenomena categorize human cognitive experience, a linguistic construction such as a construction with a clausal argument serves the purpose of indexing certain aspects of the cognitive processing of the situation which justify the choice (that is, linguistic categorization) made by the speaker in favor of one particular construction rather than an alternative one. Therefore, morphosyntactic variation within a grammatical construction reflects different cognitive paths followed by categorization processes.

Let us consider the idea of completion/incompletion of the activity expressed by the infinitive/participle constructions. For example, in (102b), *throw* is interpreted as a one-time action because the adverbial phrase *on the floor* characterizes the location of the wrapper at the moment of observation which is *after* the moment of observation of the action by the speaker. Were it not so, the locative part of the proposition 'he threw a sweet wrapper on the floor' would simply be impossible as there would be no way for the speaker to determine the intended destination of the wrapper: it could be the floor, a nearby planter, a trash can in the corner or whatever. So, at the time of utterance the action of throwing is not perceptually present in the described situation. It is this cognitive factor that leads to the choice of the infinitive, and it has nothing to do with a hypothetical ability of the infinitive to express the idea of completion of an action. This idea is the result of an inference made by the speaker: "I saw him propel a sweet wrapper in space, now the wrapper is motionless on the floor away from the thrower, therefore the action of throwing is completed". Basically, the final linguistic product we deal with in this case is not a direct reflection of our perceptual experience, it is a *reflection of our cognitive processing of perceptual data categorized in certain morphosyntactic structures.*

That this is indeed so can be seen when the participle *throwing* is substituted for the infinitive *throw*; the resulting construction, according to informants, becomes strange or doubtful:

(112) ? *I noticed him throwing a sweet wrapper on the floor, so I asked him to pick it up.*

Such "strangeness" is due to the semantic conflict between the two parts of the sentence. While the clause *I asked him to pick it up* makes sense in the case of the sweet wrapper being on the floor (otherwise, there is nothing to pick up), the construction *I noticed him throwing a sweet wrapper,* due to the categorical meaning of the participle discussed in Chapter 6, highlights the simultaneity of the act of observation and the act of throwing, thus creating an inferential gap (incoherence) between the two propositions. So, the choice made by the speaker does not depend on the structure of the action of throwing itself; it is based on the processing of the observed situation whereby the speaker's background (structural) knowledge of the world serves as a kind of *matrix* against which the new (phenomenological) knowledge is mapped. Perceptual data in the two situations, described by (102b) and (112), are categorized differently depending on the structure of the situation and its cognitive processing, this difference being reflected in the construction variation. A situation structure is a set of certain objects at a certain place at a certain time categorized by the speaker in linguistic form and organized into a meaningful unity based on the speaker's perception of, and knowledge about, the relationships between them.

Now let us look at another example. What would happen if the structures of the situations described in (101 a) and (101b) were changed? Let us imagine that it wasn't just one door that made a specific noise, but a number of doors:

(113) a. *I heard the doors slamming all night long.*

b. ?? *I heard the doors slam just after midnight.*

Although it is possible to imagine a number of doors repeatedly slamming because of a lot of comings and goings, or because of a night storm playing with the unlocked ones (113a), our experience of the world and knowledge of the way things are tell us that it is highly unlikely for a number of people, and practically impossible for a storm no matter how strong, to simultaneously slam a number of doors shut (113b). Of course, from the point of view of the hardened generativist sentence (113b) is well-formed. But then, this is where the difference between the cognitive and generativist accounts of grammatical phenomena comes to the fore: in cognitive grammar they are viewed as a reflection of cognitive processing (categorization and structuring) of perceptual data, while in generative grammar they are viewed as mere operations on symbols; the possibility and character of such operations are predetermined by the qualities of the symbols themselves without any crucial reference to the external world.

As a consequence, there is a tradition in generative grammar to take isolated sentences and analyze them as if they were existentially independent of anything else, first and foremost, of the real world situation in which they are produced by the speaker, that is, independent of the *physical context*. Constructions with non-finite clausal arguments are no exception. Consider the following nearly identical narrative samples:

(114) a. [1]*Finally, they found what they had been looking for — an opening that appeared to once have been an entrance to the mine.* [2]*They decided that Bret would go in to explore, and John would stay outside and wait for a signal.* [3]*After Bret disappeared in the dark, John nestled himself into a crevice behind a thick growth near the entrance, merging with the patchy green of the hillside and preparing to wait.* [4]*However, it wasn't before too long when he heard Bret* ***whistle/? whistling*** *softly, and he cautiously followed him into the darkness...*

b. [1]*Finally, they found what they had been looking for — an opening that appeared to once have been an entrance to the mine.* [2]*They decided that Bret would go in to explore, and John would stay outside and wait for a signal.* [3]*After Bret disappeared in the dark, John nestled himself into a crevice behind a thick growth near the entrance, merging with the patchy green of the hillside and preparing to wait.* [4]*However, it wasn't before too long when he heard someone* ***? whistle/whistling*** *and a rock or two* ***? roll/rolling*** *down the slope not too far away...*

As confirmed by informants, in (114a) the infinitive seems to be the right choice, while in (114b) the participle seems more appropriate. What is the difference between the two passages in terms of their orientational effect on the reader ('knowledge representation', in traditional terminology) that accounts for the choice of a specific non-finite verb? They both give an account of a certain train of events at a certain place during a certain time period which included an acoustic event 'whistling'. Since everything said is said by an observer, and since the author of the above narratives is not the actual observer of the situation described, for the text to have an orientational effect on the reader, the writer's source of information must be identified. Whereas the first two sentences in either passage are not suggestive of who the observer might be, the third and, particularly, the fourth sentence, whose structure is similar to the structure of the complex sentence analyzed in the first part of this chapter, give us a clear idea who the original source of information is because the situation is described as viewed and perceived by John. However, in (114a) 'whistling' is not the last event in a series, it is followed by 'John cautiously following Bret into the darkness', and this is an iconic reflection both of the chain of events as they were processed by John as the observer, and of their successive linguistic categorization by John as the conjectured speaker to whom the author owes the information shared with the reader.

In the second passage, two aspects of linguistic categorization of the acoustic events 'whistling' and 'rolling', namely, the morphological structure of the words used to refer to these events (which are present participles) and the syntactic structure of the clausal argument as a whole, cue the cognitive processing of the described situation as if it were perceived by John, i. e., it is categorized as John's *direct perceptual experience*. The reader is placed in the 'here-and-now' of John as the observer, he is given a vantage point by skipping the mediation link in the form of a conjectured speaker as a prerequisite for 'knowledge transfer' in the first passage. So, the difference between the two passages that affects the choice of a particular type of clausal argument in each case may be explained in terms of *different cognitive structures* underlying the last sentence in each of the passages. In (114a) we are dealing with *structural* knowledge; to the writer, the fact that the action of 'whistling' really took place in the sequence of described events is knowledge handed him, as it were, ready-made by a second party, that is, John, who is projected as the observer/speaker. In (114b) we are dealing with *phenomenological* knowledge; by choosing the participle, the writer cues the reader's interpretation of the text, providing him with symbolically simulated 'direct perceptual access' to the described situation, thus orienting him in his interaction with the text as a symbolized sign system for storing knowledge.

7.2.3. Clausal Arguments as Structures for Knowledge Representation

As was argued in Chapter 6, the infinitive does not refer to something which exists in the real world as it is perceived by an observer. Rather, and this must be especially emphasized, it refers to an abstract concept which does not possess any spatio-temporal ties with perceived reality, the referent of the infinitive cannot be a concrete perceivable entity/phenomenon in principle. Only that is accessible to our senses which is part of our perceived world, just like any part of the perceived world is conceived as such only because it can be directly accessed by the senses, that is, on the level of phenomenology. By contrast, the referent of the (present) participle is part of the world by definition, therefore any act of reference in the case of the present participle implies *observability* and/or being in the domain of shared current experience. Any claims that thereby the object of perception is some middle stage in the unfolding process or ongoing activity are nothing but conjectures not sustained by empirical evidence.

Suppose I am walking down the street and happen to glance at an open window and notice there a couple moving in what appears to be a dancing *pas* — what grounds do I have to claim that they are in the middle (beginning, end) of the dance? It might be the first movement in a dance just begun, or it might be the final movement after which the dance dies, leaving no trace in the observed space-time. However, in either case I have a right to say:

(115) *I saw them dancing*

for it was dancing that I saw. What kind of dance was it? Was it danced from beginning to end or did the partners go through just a few movements on the spur of the emotional moment — who knows? But one thing I know for sure: at the moment I cast my glance on a certain spatial domain constituted by certain objects, it featured, existentially, what in English is called "dancing".

It is not that simple when we turn to the infinitive construction. The difficulty is caused not so much by some mysterious properties of the infinitive, but rather by the specific properties of the main verb introducing the construction. As has been mentioned, these are usually perception verbs, also known as cognitive verbs because they identify the source of information expressed in the proposition that follows. As a matter of fact, when I say:

(116) *I saw them dance*

I am saying that I know something indicating, at the same time, how I came by this knowledge. The meaning of such an utterance is as follows: "I know that in the world as it has been experienced by me *they* at least on one occasion entered in a certain spatio-temporal relation called "dance" and, therefore, I have reasons to believe that they knew how to do it and could do it". There isn't any implication in this statement that the activity in question was observed by me from beginning to end in its entirety. Likewise, when I say:

(117) *I heard her sing*

in no way do I make the claim that I listened through the piece from beginning to end. The meaning of such an utterance is "I-here-now know that she-there-then could sing because I heard her". The meaning of (118) is something like "I-here-now refer to I-there-then at the place and time where she-there-then was singing":

(118) *I heard her singing.*

Now let us look at the behavior of perception verbs and verb phrases which are not cognitive in the strict sense of the word. They allow only the participle construction and reject the infinitive:

(119) a. *I* detected *him in the lobby unobtrusively* making *his way into the porters quarter.*

b. **I* detected *him in the lobby unobtrusively* make *his way into the porters quarter.*

(120) a. *I* caught *the kid* stuffing *his pockets with fistfuls of cheap candy.*

b. **I* caught *the kid* stuff *his pockets with fistfuls of cheap candy.*

(121) a. *Finally we* spotted *him* crossing *the street at the far end.*

b. **Finally we* spotted *him* cross *the street at the far end.*

Why this constraint? Apparently, it is imposed by the semantics of the main verb. Part of the meaning of the verbs in (119) – (121) is what may be described as 'momentary awareness of an entity's existence caused by perception'. And, as I have argued, it is just this feature that distinguishes the referent of the present participle from the referent of the infinitive. Verbs like *detect* or *spot* are not 'cognitive' for the reason that their function as predicates is limited by the spatio-temporal boundaries of the moment at which perceptive awareness of an entity's existence occurs. Conversely, the function of 'cognitive' verbs as predicates is not related to any limit imposed by the spatio-temporal grid of the situation as the propositional content: such verbs introduce a proposition as a quantum of interiorized knowledge which, in a way, is analogous to the systemic meaning of the infinitive as specified above. The meaning of (116) as opposed to the meaning of (115) is characterized by two hierarchically related planes: on the one hand, and this is meaning number one, I claim that they were able (could, knew how) to dance, and this knowledge is part of my conception of the world as I know it. On the other hand, and this is meaning number two, I give an argument for the sufficiency of grounds on which the claim is made by indicating the source of my information ("I saw it"). In other words, a construction with the infinitive clausal argument is a means of linguistic representation of s t r u c t u r a l knowledge of the world.

In the case of participle construction, meaning number one is absent altogether because the proposition expressed in the clausal argument is purely existential in the sense that its truth conditions are limited only by the temporal frame of the observer's perception of the described situation ('they dancing'), that is, the proposition is essentially phenomenological. But if this is the case, as soon as the temporal frame boundaries have been trespassed, the inclusion of this proposition in the general worldview in the guise of an interiorized knowledge structure independent of specific phenomenological manifestations of the world falls out of the question.

Unlike cognitive verbs, causative verbs in constructions with clausal arguments always take the infinitive and never the participle:

(122) a. *They* made *him* leave *the room.*

b. **They made him* leaving *the room.*

(123) a. *She* let *me* do *it my way.*

b. **She* let *me* doing *it my way.*

(124) a. *He* helped *her* pack *the stuff.*

b. * *He* helped *her* packing *the stuff.*

Such selectivity may, again, be put down to the specific meaning of the main verb: causation is directed toward a desirable, rather than existing, state of affairs, so a linguistic sign used to express a desirable state of affairs cannot refer to any part of the real world (i. e., the world as perceived by the observer). Hence the constraint on the participle in constructions with causative verbs. For similar reasons, many other verbs that may be followed by a clausal argument allow only the infinitive and never the participle. For example, 'query' verbs, such as *ask*, refer to an act of verbal behavior aimed at soliciting a behavioral response from a protagonist; such response can only be expected or anticipated, so, from the point of view of the languaging observer, it is not, and may not be, part of the perceptually processed and linguistically categorized real-world situation:

(125) a. *I asked John to help.*

b. **I asked John helping.*

(126) a. *She implored him to go.*

b. **She implored him going.*

7.3. Summary

Building on an autopoietic approach to language as a special dimension of the cognitive domain of interactions, and on phenomenological nature of linguistic signs which are grounded in our experience of interactions in a consensual domain, I argued in Chapter 5 for a revised conception of grammar. Rather than providing rules for combining and recombining linguistic forms to express meanings as 'truths' about the 'real world', grammar is a means for categorizing patterned recursive experience of both linguistic and extralinguistic phenomena, or knowledge. Knowledge, therefore, is understood as live action described by a languaging observer as a function of fine structural coupling of an organism with the environment, and as 'meaning-making' it lies at the core of the cognitive science enterprise. 'Chunks' of knowledge-as-live-action are represented in grammatical categories as metasigns, and in spoken language the function of such metasigns is to construe and organize the world as it is 'described' by our linguistic behavior in a consensual domain. Such construal and organization are effected by partitioning the world with the help of linguistic signs grounded in first-order activity.

In written language as a cultural artifact, grammar is a symbolized sign system for storing knowledge; in this system, every grammatical category, or metasign, is part of a cognitive structure, such as a phrase or a sentence, and relates to some essential aspect of cognitive processing in a first-order consensual domain. By taking two types of linguistic structures, a complex sentence and a construction with clausal arguments, I have shown how the semiotic aspects of grammatical categories as metasigns, particularly, iconicity and indexicality, interact in the second-order consensual domain, making knowledge representation and its understanding possible. Viewed from this angle, the structure of a complex sentence of the type *When S, S* and its possible variations turned out to depend on an interplay of such factors as the actual order of the clauses, the relative position of the indexical *when,* the kind of aspect expressed by the verb, and the presence or absence of a verb complement in a clause. All these structural features cue understanding by orienting the listener or reader in the second-order consensual domain depending on his individual experience of interactions in a first-order consensual domain, where the initial grounding of linguistic signs takes place.

A methodologically consistent approach to the problem of clausal argument variation in constructions of the type *X saw Y do/doing smth* allowed to relate it to the categorical grammatical meaning of the infinitive and participle as described in Chapter 6. This meaning, expressed by the grammatical category of aspect, is indication to the source of information about the event categorized by the non-finite verb. Thus, the two types of clausal arguments are used to categorize two types of knowledge — phenomenological, in the case of participial construction *X saw Y do smth*, and structural in the case of infinitival construction *X saw Y doing smth*.

The analysis undertaken in this chapter contributes to the general effort of cognitive science to resolve the problem of knowledge representation by setting specific guidelines for similar analyses of various linguistic phenomena, from morphology to syntax, by highlighting the functional properties of linguistic signs and metasigns as empirical objects integrated in the dynamically complex behavior of cognizing humans as unities of interactions (second-order living systems) on the one hand, and the role of grammar in categorizing experience of such interactions, on the other.

CHAPTER 8

NOMINAL GENDER: A CASE FROM RUSSIAN

8.1. The Problem with Traditional Semantic Classification

In Russian, the category of gender is a grammatical property of pronouns, nouns, adjectives, and verbs in the past tense. Adjectives are marked for gender to agree with nouns, and verbs in the past tense agree with nouns and pronouns. Pronouns and nouns, therefore, are of primary interest to those studying gender as a grammatical category.

If we compare the ways gender is expressed in nouns and pronouns, we will find that they are basically inconsistent, but inconsistent in a different manner. Pronouns are marked for gender in the 3 person singular and are unmarked in the 1 and 2 person singular; with the exception of Semitic languages, this seems to be a universal feature across languages. It stems from the epistemic mode in which the world is categorized by means of a tripartite classification of entities into the subject of speech, the object of speech, and the subject matter of speech (Kravchenko 1992). To clarify the meaning of the latter statement, let us look at the etymology of the Latin word *persōna*. Although *Oxford Latin Dictionary* (Glare 1982) suggests that it may be a borrowing from Etruscan, its relationship to the Gk. *πρόσωπον* is obvious as it is the loan translation of the latter. Both the Latin and Greek words have the same original meaning 'actor's mask; role in a play or a tale'. The noun *persōna* correlates with the verb *personō* 'to make a loud, continuous sound' which, in its turn, is related to the noun *sonus* 'sound, noise' with the prefix *per-* which is derived from the preposition *per* with the core meaning 'through, by' (that is, it points to the medium through which an object is perceived). The etymology of the Gk. *πρόσωπον* is less clear; however, if the prefix *πρό-* descends from the preposition *πρό* 'before', which seems very likely, *σωπον* may be related, for example, to the verb *σωπαινω* 'to stop speaking, become silent'. It may be assumed that the meaning of both the Latin and the Greek word involved the concept of a source of continuous sound, namely, the flow of speech which could be a classificatory criterion in dividing all objects into speaking and non-speaking. For example, according to Claude Levi-Straus (1962), such classification is found in the language of Navahos, the principal tribe of the southern division of the Athapaskan stock of North American Indians: they divide all living creatures into two categories with regard to their speech ability.

The primary epistemic contrast "I - you" reflects the first and most important step in categorization which is the phenomenological core of language as a patterned system of coordinated orientational activity. The phenomenological nature of this epistemic contrast renders explicit gender identification irrelevant since such identification is not an issue in a canonical situation of utterance. Third person pronouns, on the contrary, reflect the twofold differentiation of objects in the 'real world' on a primary feature "animate/inanimate", and a secondary feature of gender. This classification is, essentially, pragmatic in that it is predetermined by the two existential imperatives for humans as living systems: survival (an animate object is a high-ranking source of danger), and procreation.

According to standard academic grammars, most nouns in Russian are marked for gender, with the exception of *pluralia tantum* nouns and the so-called "common gender nouns" such as *sirota* 'orphan', *umnitsa* 'smart one', *rabotjaga* 'laborer', etc. (Shvedova and Lopatin 1989). To quote an authoritative source,

> the category of gender of nouns is a non-formative syntagmatically revealed morphological category manifested in the ability of singular nouns to relate preferentially to the gender forms of those word forms that agree (or are coordinated, in the case of predicates) with the noun (Shvedova 1980, 1: 465).

The definition quoted above, although it accurately describes the *manifestation* of gender as a morphological category, fails to give any rationale behind the classification of nouns into different gender classes; therefore, the implication is that the classification itself is arbitrary, which stands well with the structuralist 'arbitrariness-of-sign principle'. To say that nominal gender is a 'syntagmatically revealed morphological category', that is, that a speaker of Russian learns about the gender class of a particular noun from 'those word forms that agree (or are coordinated, in the case of predicates) with the noun' is, so to speak, "to put the cart before the horse": it is just because adjectives or verbs agree, or are coordinated with, nouns that they *must* take inflections which mark the gender class of the noun they agree or are coordinated with. Thus, in order to be able to say something like *star<u>yj</u> lar'* 'old-*masc.* chest-*masc.*' or *star<u>aja</u> gar'* 'old-*fem.* burnt-out forest-*fem.*', where the adjective "old" is marked for masculine gender in the first case and for feminine in the second, the speaker must know that *lar'* is masculine, and *gar'* – feminine, even though there is absolutely nothing in the shape of the words themselves to indicate the difference. So we are dealing here with tacit knowledge, and if knowledge is categorized experience, then the seemingly unexplainable gender classification of Russian nouns could hardly be irrational (cf. Lakoff 1987).

True, inclusion of nouns in a specific gender class does seem to be arbitrary both on grammatical (morphological) and semantic criteria. Whereas in the case of 3 person pronouns the grammatical marker for gender is morphologically consistent (Rus. *on* 'he', ***ona*** 'she', ***ono*** 'it'), nouns defy any such consistency. In Russian, alongside with the typical morphemes ***-a***, ***-ja*** as in *zhen**a*** 'woman/wife', *zeml**ja*** 'earth', feminine gender can be expressed by the so-called zero morpheme as in *kost'* 'bone', *plot'* 'flesh', *shal'* 'shawl, *sazhen'* 'arch. linear measure unit' typical for masculine nouns such as *gost'* 'guest', *tat'* 'arch. thief', *pleten'* 'wicker fence'. Likewise, the ***-a, -ja*** morphemes, which typically 'mark' nouns as feminine, may be found in masculine nouns (*muzhchin**a*** 'man', *pap**a*** 'dad', *djad**ja*** 'uncle'). The neuter gender is typically marked by the ***-o*** and ***-je*** morphemes, but the morpheme ***-ja*** is not uncommon in neuter nouns such as *dit**ja*** 'child', *plam**ja*** 'flame', while the ***-o*** and ***-je*** morphemes are found in masculine nouns as well, such as the diminutive *domishk**o*** 'little house', *golosishch**e*** 'loud booming voice', *podmaster**je*** 'apprentice'. As data indicate, the grammatical category of nominal gender seems to reflect a rather arbitrary classification, since there isn't a consistently applied formal feature that would unambiguously identify the gender of a given noun.

In Russian, morphological gender markers are often tied to the word formation (derivational) processes. Remarkably, the masculine noun appears to be the primary form, as the word formation process follows the direction "masculine —> feminine", similar to the pronominal paradigm *on* 'he' —> *ona* 'she':

(127)	BASE STEM (MASCULINE)		DERIVED STEM (FEMININE)
	lis 'male fox'	—>	*lisa* 'female fox'
	volk 'male wolf'	—>	*volchitsa* 'female wolf'
	uchitel' 'male teacher'	—>	*uchitel'nitsa* 'female teacher'
	povar 'male cook'	—>	*povarixa* 'female cook'
	sneg 'snow'	—>	*snezhinka* 'snowflake'
	dozhd' 'rain'	—>	*dozhdinka* 'raindrop',
	etc.		

In Indo-European, the ***o***-suffix was used to refer to male objects, whereas female objects were marked by the ***a***-suffix and ***-ie*** , ***-i-***. According to Karl Brugmann (1889), at first none of these suffixes had anything to do with sex distinctions. He hypothesized that, probably, the idea of female nature inherent in the root of one or two words — possibly, **genā* 'wife (= woman)', **mămā* 'mother' — was transferred onto the suffix which began to be used to differentiate between male and female species. It was only due to the contrast with ***-i-*** that the ***-o-*** form began later on to be used to refer to male species, and it always remained the primary means of naming any animate creature without indicating gender distinctions.

It is difficult to find semantic justifications for classifying nouns into different gender classes. If, in the case of animate nouns, their gender can be explained through gender distinctions of respective referents, it cannot be done so in the case of inanimate objects. Why are *den'* 'day' and *vecher* 'evening' masculine, *noch'* 'night' — feminine, and *utro* 'morning' — neuter? Why are *pen'* 'tree stump', *kisten'* 'flail' (as a weapon), *lar'* 'chest' masculine, and *ten'* 'shade', *sazhen'* '*arch.* linear measure', *gar'* 'burnt out (place in a) forest' — feminine? Why is *ryba* 'fish' feminine, while the names of most of the common kinds of fresh-water fish such as *okun'* 'perch', *som* 'cat-fish', *ugor'* 'eel', *jorsh* 'ruff', *karp* 'carp', *peskar'* 'gudgeon', etc. are masculine? Why is *dub* 'oak' a "he", and *berjoza* 'birch tree' — a "she"? Is there any logic at all in how Russian nouns are classified into different gender groups?

As has been mentioned above, feminine nouns are typically derived from masculine nouns. This parallels the morphological pattern for 3 person pronouns, so the following question seems to be not entirely out of place: "Is it a mere coincidence, or may it be an indication that some profound process is at work here, and certain cognitive mechanisms determine the way gender distinctions are categorized in nouns?" The answer may lie in establishing the relationship between nouns and pronouns as two basic classification systems which natural language uses to partition the observed world.

8.2. Nouns vs. Pronouns

8.2.1.The Logic of Naïve Classification

In orthodox linguistics, whose conceptual inventory has been much affected by ill-applied common sense, definitions of pronouns based on their allegedly substitutive function (i. e., when they are viewed as replacements for nouns or noun phrases) have become axioms *sine qua non.* Two factors have played a role in the establishment of this axiomaticity: 1) the conventional interpretation of the meaning of the Lat. *prōnōmen* believed to be "[that which is] instead of noun", and 2) their frequent anaphoric use *in written language* when a pronoun is preceded by a noun phrase (antecedent). As a consequence of such assumptions, pronouns are said to be used in discourse referentially only if there is a nominal antecedent because they have little or no intrinsic meaning or reference. This is just another example of how identification of writing (texts) with natural language (discourse) leads to a distorted picture orthodox linguistics draws of linguistic signs and their function, as in the case of pronouns. Naturally, the view that pronouns are substitutes for nouns which must precede pronouns to make them referential, is absolutely inadequate.

As a system of signs used by humans to categorize experience, the lexicon of natural language may be divided into several subsystems based on particular classificatory criteria humans are guided by in partitioning the perceived and enacted world. As part of the lexicon, nouns and pronouns are just two such classification systems. Prototypically, nouns categorize 'thingness' as a property of objects which is related to their ability to preserve the same volume in space (a spatial domain) they occupy over time. In other words, categorization of an entity as a 'thing' directly depends on our perception of three-dimensional space (Vaina 1990). The world, as a totality of potentially perceived things and phenomena, is infinite, therefore, the number of spatial domains that may potentially be categorized as 'things' is also infinite in principle, which accounts for the well-known fact that the nominal class in any given language is an open set. By contrast, pronominal classes in different languages are closed sets of signs of particular type. Thus, so-called 'personal' pronouns are lexical items used to categorize three types of entities with reference to the speaker: 1 person pronoun points to the speaker, 2 person — to the person spoken to (addressee), and 3 person — to a person or thing that are neither the speaker nor the addressee (Jespersen 1929; Lyons 1968). This classification is universal for different languages (Forchheimer 1953; Trask 1993).

So, on the one hand, we have a 'primitive' and ingenious in its simplicity classification of all that is into three categories: "I", "YOU", and "EVERYTHING ELSE". The class of objects that constitute the first category is a set always equal to unity (the difference between the singular and the plural, which receives a formal distinction in language, may be neglected here as a conceptual extension: *we* = 'I and whoever I associate myself with at the moment'). The class of objects constitutive of the second category is also equal to unity in prototypical acts of communication, and although it may present a set greater than unity, it always remains a closed set of objects that are, perceptually, within an earshot (again, we are speaking here of *natural spoken* language). However, the class of objects constitutive of the third category is an open, potentially infinite set of (again, prototypically) *visually perceptible* entities.

On the other hand, we have a much more complex classification when

(i) concrete physical objects are categorized as individual holistic instances of a class, e. g., *tree, beast, house*;
(ii) groups of similar concrete objects are categorized as spatial domains on the criterion of their compact location in the space they occupy, e. g., *tree* → *forest*, (wild) *animal* → *pack* (of animals), *house* → *village*;
(iii) concrete physical objects as holistic entities are reconceptualized as structured spatial domains constituted by objects, e. g., *tree* → *trunk, root, top, branches, leaves*; *beast* → *head, body, legs, tail; house* → *wall, roof, porch, door, window*;

(iv) within broad universal classes such as (i) objects are further categorized into subclasses of concrete things possessing a set of permanent secondary distinctive features, e. g., *tree* → *pine, fir, birch, aspen; animal* → *bear, wolf, fox, hare; house* → *cabin, cottage, hut.*

The exact number of criteria used in such classifications is hard to define, but then, this is not our goal. Of real interest is the question whether there are any grounds to assume that a complex classification based on abstract concepts is, in fact, epistemologically *primary* to a simple classification based on purely phenomenological criteria — for this is precisely the difference between the nominal and pronominal classification systems in natural language. The strong belief shared by orthodox linguists that pronouns substitute nouns leads to an obvious inference that an intricate, finely tuned classification system as represented by nouns, is primary to a crude unsophisticated classification system represented by pronouns — which runs counter to what modern science tells us about cognition.

8.2.2. Etymology

The meaning of the Lat. *prōnōmen* is usually interpreted as 'the word used instead of the noun proper' (Glare 1982). The original meaning of the expression *instead of* is 'in place of', but, because of the written language bias, it has always been treated as a synonym to the expression *in exchange for*: in writing, for a pronoun to be used referentially, it must, typically, have an antecedent, so the orthodox argument is that the role of the pronoun is to help avoid naming the same referent twice by simply pointing at it (the anaphoric function of pronouns). This argument has been more than once refuted in the literature (Quine 1961; Geach 1962; Thavenius 1983), so we are not going to be concerned about it.

The Lat. *prōnōmen* is traced to the prepositional construction *prōnōmine* (Klein 1967), while one of the meanings of the preposition *prō* is 'priority in space or time' which is well illustrated by motion verbs with the *prō-* prefix, which is derived from the preposition *prō*:

(128) *<u>prō</u>venio* 'move forward',

<u>prō</u>trudō 'project, thrust forward',

<u>prō</u>tendō 'spread before oneself', etc.

Compare also *proavctor* 'forefather, ancestor', *proavia* 'great-grandmother' and the like: a forefather is not someone who substitutes another, more recent ancestor, neither can it be said about a great-grandmother that she somehow substitutes a grandmother. Words like *proavctor*, *proavia* refer to those who were before in time, so when the function of personal pronouns comes into focus, the formulation 'instead of a noun' should not be interpreted in the sense that first there was a noun which later was replaced or substituted by a pronoun, but in the sense that the position of pronoun in the sentence is the position assigned by language as it is today to nouns. Thus it becomes apparent that the logic of naïve classification and the etymology of the word *pronoun* speak in favor of the ontogenetic primacy of pronouns, supporting the claims made in early 20th century by Karl Brugmann (1904).[26] A pronoun is not something used instead of a noun, it is what precedes it in space and time, that is, a *protonoun*.

With this in mind, the view often expressed in the literature that the linguistic function of pronouns is to substitute nouns and that, as a consequence, they possess a highly generalized meaning, seems unjustified. Common to all pronouns as a class of lexical items is that they are indexical signs (Kravchenko 1992). In a prototypical situation of oral communication we point verbally to perceivable objects which are in our perceptual space and thus accessible to perception (Bennett 1978; Huddleston 1984). The singling out of a thing from among other things by means of indication is, epistemically, an act of primary categorization based on pure phenomenology. Generalization, however, is a higher-order cognitive activity involving transition from narrower-scope concepts to concepts of a wider scope. The naming of a thing, hanging a 'label' on it (Geach 1962) as the linguistic function of nouns, cannot precede phenomenological identification of the object to be named, or its singling out as an existing entity via indication, which is the linguistic function of pronouns. This idea was succinctly expressed by Charles Peirce in his *Elements of Logic*:

> There is no reason for saying that *I, thou, that, this,* stand in place of nouns; they indicate things in the directest possible way. It is impossible to express what an assertion refers to except by means of an index. A noun, on the other hand, does not *indicate* the object it denotes; and when a noun is used to show what one is talking about, the experience of the hearer is relied upon to make up for the incapacity of the noun for doing what the pronoun does at once. Thus, a noun is an imperfect substitute for a pronoun (Peirce 1932: 163).

8.3. The Hypothesis

If we assume that pronouns have ontogenetic primacy over nouns, then it can be hypothesized that gender differentiation of pronouns was the next step in creating the taxonomic model of the world, the first step being categorization of all phenomenologically accessible reality into two realms: participants in the speech act

---

[26] It must be noted that this view was strongly opposed by Karl Bühler (1934).

(in a canonical situation of utterance, the speaker and the addressee) and the rest of the world (Kravchenko 1992). This, basically, made the epistemic contrast "I-you" the ultimate point of reference in natural language, all possible referents being subsumed by the concept "it" (= 'the subject matter of speech') as anything that is not immediately embraced by the first concept.

It can be further hypothesized that in the course of time non-gender specific identification of the subject matter of speech (*ono* 'it') gave way to the contrast *on* - *ono* ('he - it') on the 'animate/inanimate' feature. Circumstantial evidence to this is found in the fact that, typically, nouns denoting animate creatures (names of species) that can be considered as prototypes for respective classes, are masculine:

(129) <u>Masculine Nouns as Names of Species</u>

volk	'wolf',
zajats	'hare',
medved'	'bear',
orjol	'eagle',
golub'	'pigeon',
vorobej	'sparrow',
karp	'carp',
okun'	'perch', etc.

The animalistic past of language also helps explain why many inanimate nouns, especially names of natural phenomena such as *veter* 'wind', *grom* 'thunder', *dozhd'* 'rain', *grad* 'hail', *sneg* 'snow', *inej* 'hoarfrost', *moroz* 'frost', *ogon'* 'fire', are masculine: as David Premack hypothesizes, the attributing of the feature "animate" to an object in the case of an infant depends on whether that object belongs to the self-propelled or nonself-propelled class.

> Self-propelled objects are those objects that can both move and stop moving without assistance from another object; nonself-propelled objects are those objects that cannot. Motion *per se* is not a critical parameter. <...> Change is what is critical — from rest to motion (or vice versa), one speed to another, one direction to another — and the ability of an object to execute these transitions without assistance from another object (Premack 1990: 3).

Premack suggests that induced changes in movement in nonself-propelled objects are perceived by the infant as causal, while changes in the movement of self-propelled objects (and, perhaps, changes in color, shape, size, or temperature) are perceived as intentional: "Just as causality is the infant's principal hard-wired perception for nonself-propelled objects, so intention is its principal hard-wired perception for self-propelled objects" (*ibid.*, p. 3). Thus the concepts of causality and intentionality, which are crucial in understanding cognition as a life process (see discussion in Chapter 3), appear to be directly linked to the perceptual experience of the observer. This experience, for example, told the naïve observer of the long ago that the falling of snowflakes was not something induced by another object, therefore snow was a self-propelled, i. e., intentional (animate) object.

At first, it seems, the contrast *on* - *ona* 'he - she' reflected natural gender distinctions of humans. Then, it was extended to other animate objects and, finally, to any obviously inanimate objects, the extension being based on analogy and functional association. The analogy principle can be traced in the gender characteristics of some nouns that are names of natural objects, when the prototypical features that distinguish male from female can be attributed to those objects, as in Table 12.

Table 12. Nominal gender by analogy with cognitive prototypes

Gender	**Noun**	**Meaning**	**Prototype**	**Feature(s)**
Masculine	*dub*	oak	Male	big, strong, sturdy
	ogon'	fire		powerful, dangerous
	kamen'	rock		hard, rigid, unyielding,
Feminine	*rjabina*	mountain ash	Female	slim, thin, yielding
	voda	water		soft, caressing, soothing
	zemlja	earth		soft, tender, life-giving

The functional association principle in noun gender classification may be found in groups of names of certain artifacts: *nozh* 'knife, *m*', *topor* 'ax, *m*' — implements used by man as provider and protector, *lozhka* 'spoon, *f*', *ploshka* 'earthen saucer, *f*', *chashka* 'cup, *f*' — implements used by woman as a hearth keeper; *luk* 'bow, *m*' vs. *strela* 'arrow, *f*' — dependence relationship (hierarchical subordination); *koren'* 'root, *m*', *stvol* 'bole, *m*' vs. *krona* 'tree-top', *vetka* 'branch, *f*' — one as the foundation and support for the other. Numerous examples of this kind make it possible to hypothesize that the category of gender in Russian reflects the workings of a certain system of classifiers that are cognitive in nature and structured around a secondary epistemic contrast *on*/a – *ono* ('he/she – it'), i. e. *the grammatical category of noun gender is a reflection of the basic world classification as represented by the system of personal pronouns.*

8.4. Case Study

A case study of gender characteristics of names of animals in Russian seems to provide supportive data to this hypothesis. As may be seen from Table 13, animals common in the geographic area historically populated by Eastern Slavs can be divided into three main groups: 1) animals dangerous for man, 2) animals useful for man, with two subgroups — (a) source of food, (b) source of warm hide/fur, 3) other animals (of no particular practical interest).

Nouns that are names of species in a scientific classification belong, as a rule, to the masculine class. In the predator group, however, two nouns, which are names of species, belong to the feminine class: *rys'* 'lynx', *lisa* [*lisitsa*] 'fox'. Although the noun *rys'* does not have an opposite gender correlate, the nouns *lisa, lisitsa* are derived from the masculine noun *lis*. In the scientific classification, it is the feminine noun *lisitsa* that is used as a name of species. It is derived from the masculine noun on the same pattern as the feminine nouns *medveditsa* 'she-bear', *volchitsa* 'she-wolf', whereas the meaning 'female fox' is represented by the form *lisa*, which is also used as the name of species in the naïve classification. How can such inconsistency be accounted for?

If we assume, following Derek Bickerton (1990), that words and concepts in natural language are generated by functional utility, and agree with George Lakoff (1987) that our bodily experience and the way we use imaginative mechanisms are central to how we construct categories to make sense of experience, the above inconsistency may be explained by a certain naïve set of classifiers, such as follows:

1) the size of the animal (big vs. not big),
2) immediate danger to man (animal, an encounter with which is life threatening vs. animal, an encounter with which is not life threatening),
3) edibility (edible vs. inedible),
4) possibility to use animal hide for warm clothing (furry vs. non-furry animal),
5) visually distinguishable gender of the animal.

As Table 13 shows, gender differences among names of predators in Group 1 reflect consistent differences among their referents on the first and second classificatory features: bears and wolves are relatively big in size and are reputed (at least, in folklore) man-eaters, while foxes and lynxes are of noticeably smaller size and, respectively, of lesser strength, so they do not stalk humans. It is worthy of note that this differentiation on the feature "dangerous" is analogous to that between man and woman in the primeval society where the main social functions enjoyed by men were the providing of food and the waging of war, and they both involved killing live creatures.

Table 13. The naïve classification of wild animals

	Name of species (with fem. nouns in italics)	Meaning	Opposite gender correlate	Classificatory feature				
				Relatively big in size	Immediate danger	Food	Hide/fur used for clothing	Visually distinguish-able gender
1	medved'	bear	medveditsa	+	+	+	+	—
	volk	wolf	volchitsa	+	+	—	+	—
	lisa, lisitsa	fox	lis	(+)	(+)	—	+	—
	rys'	lynx	—	(+)	(+)	—	+	—
2	zubr	aurochs	zubrixa	+	(+)	+	+	+
	kaban	wild boar	kabanixa	+	(+)	+	(+)	+
	olen'	red deer stag	olenuxa	+	—	+	+	+
	izubr'	elk	—	+	—	+	+	+
	los'	moose	losixa	+	—	+	+	+
	kozjol	goat	koza	(+)	—	+	+	+
	lan'	fallow deer	—	(+)	—	+	+	(+)
	kosulja	roe	—	(+)	—	+	+	(+)
	zajats	hare	zajchixa	—	—	+	+	—
3	sobol'	sable	—	—	—	—	+	—
	pesets	polar fox	—	—	—	—	+	—
	xor'	polecat	—	—	—	—	+	—
	gornostaj	ermine	—	—	—	—	+	—
	kolonok	kolinsky	—	—	—	—	+	—
	surok	marmot	—	—	—	—	+	—
	barsuk	badger	—	—	—	—	+	—
	jenot	raccoon	—	—	—	—	+	—
	bobr	beaver	—	—	—	—	+	—
	krot	mole	—	—	—	—	+	—
	vydra	otter	—	—	—	—	+	—
	kunitsa	marten	—	—	—	—	+	—
	rosomaxa	wolverine	—	—	—	—	+	—
	belka	squirrel	—	—	—	—	+	—
	norka	mink	—	—	—	—	+	—
	laska	weasel des-	—	—	—	—	+	—
	vyxuxol'	man	—	—	—	—	+	—
4	jozh	hedgehog	jezhixa	—	—	—	—	—
	burunduk	chipmunk	—	—	—	—	—	—
	xomjak	hamster	—	—	—	—	—	—
	suslik	gopher	—	—	—	—	—	—

Species names for edible animals in Group 2 are, for the most part, also masculine. Exceptions are *koza* 'goat, *f*', *lan'* 'fallow deer, *f*', *kosulja* 'roe, *f*' which are all feminine. *Kozjol* 'goat, *m*' is a species name in the scientific classification, although it is derived from *koza* 'goat, *f*' (which, like the noun *lisa* 'fox, *f*', is a species name in the naive classification). This may be due to the unification of the semantic paradigm of species names.

Morphological similarities between the Old-Slavic stems in the nouns *olen'* '(red deer) stag' and *lan'* 'fallow deer, *f*', as well as the fact that there is an etymological correlate *laně, laň* 'female deer' in Czech, indicate that, probably, *lan'* as a species name was the result of a later semantic shift in the name of female red/fallow deer. This may account for the absence of a morphological correlate with the meaning 'male' for *lan'*. Absence of a similar correlate for the noun *kosulja* 'roe, *f*' may be the consequence of the word formation pattern that gender contrasted nouns follow: noun, *m* + derivational suffix —> noun, *f* (compare with the noun *rys'* 'lynx, *f*' which does not have a gender contrasted correlate).

That species names *lan', kosulja* (alongside with *koza* as a species name in the naïve classification) belong to the feminine class may, at least partially, be explained by the fact that they denote animals considerably smaller in size than their cousins — the red deer, the elk, and the moose (compare with a similar distinction in the predator group). The fact that there are derived feminine nouns in Group 2 may be due to the possibility to visually distinguish male and female species: the tusks of a boar, the antlers of a stag or moose, and the horns of a goat are of noticeably bigger size with males, rather than females. As for the noun *zajchixa* 'female hare', it can be put down to the productive word formation pattern "masc. noun stem + fem. suffix".

In Group 3 (fur animals) more than half of the species names are also masculine (10 out of 17). The noun *norka* 'mink, *f*' is the result of metonymic transfer from the fem. noun *nora* 'burrow', so its gender characteristic does not relate to the naive system of classifiers. The species name *kunitsa* 'marten, *f*' descends from Old-Rus. *kuna* 'marten pelt worth 1 dirham' (cf. the pair *lisa - lisitsa* 'fox, *f*') which suggests that there might have existed in Old-Rus. the masculine noun **kun*.

The feminine gender of *vydra* 'otter' and *vyxuxol'* 'desman' relate to the tripartite world classification common to many so-called 'primitive' languages: **water** (fish), **air** (birds), **earth** (mammals) (Levi-Strauss 1962). If we assume that this taxonomic principle was effective in Russian, then the feminine gender of *vydra* and *vyxuxol'* would be accounted for by the fact that these animals belong to the domain "water" (i. e. the fish class, and "fish" in Russian is feminine – *ryba*). In the case of *vydra* 'otter, f' we have clear etymological evidence to support the claim, as its root is traced to the Indo-European stem **udra* 'water'. It is worthy of note that, as native speakers of Russian, little children whose world view is largely naïve rather than scientific, use the words *ryba* 'fish, *f*', *rybka* 'little fish, *f*' to refer to different aquatic creatures such as tadpoles, small crustaceans, etc., until they have been taught the 'right' names by educated adults.

The feminine gender of *rosomaxa* 'wolverine', *belka* 'squirrel' seems to be unmotivated, but only because language in its modern state does not allow us to trace the original classification principle and the rationale behind it. However, Pavel Chernyx (1994) suggests that modern Russian *rosomaxa* is the transformed Old-Russian masculine noun *rozsomok* 'evil spirit'. As for *belka*, according to the same source it is the result of substantivization of the adjective *bela* 'white, f' which in Old-Russian was part of the name *bela veveritsa* 'white squirrel, f'. In its turn, *veveritsa* is a derived suffixed feminine noun of the same morphological class as *lisitsa* 'fox, f'. Thus it appears that in Group 3, 14 names out of 17 (~ 85%) were, originally, masculine.

None of the nouns in Group 3 have opposite gender correlates registered in dictionaries, although in common speech feminine nouns are easily created on the spur of the moment by adding suffixes and endings to masculine noun stems:

(130)

Masculine Stem →	Feminine Noun
sobol'	*sobolixa* 'female sable'
surok	*surchixa* 'female marmot'
barsuk	*barsuchixa* 'female badger',
bobr	*bobrixa* 'female beaver',
krot	*krotixa* 'female mole'.

At the same time, feminine nouns do not yield masculine correlates as might be expected, and the following forms are not allowed: **vydr* 'male otter', **kun* 'male marten', **rosomax* 'male wolverine', **belk* 'male squirrel', **nork* 'male mink'.

In Group 4 (other animals) all the species names are masculine, and only one has an opposite gender correlate registered in the dictionary: *jozh* 'male hedgehog' – *jezhixa* 'female hedgehog'.

Now, why are some feminine nouns that can be derived from masculine nouns on a productive pattern, registered in dictionaries (e. g. *medveditsa* 'female bear', *volchitsa* 'female wolf', *olenuxa* 'female red deer'), while others (such as *barsuchixa* 'female badger', *bobrixa* 'female beaver', etc.) are not? Apparently, such classification inconsistency is rooted in human experience, that is, in the possibility for female animals followed by their young to be more or less frequently observed by man. Since in the old times hunting was the main source of food, man had a lot of opportunities to observe wild boar or deer with their young. On the contrary, animals from Group 3 and Group 4 live mostly in burrows, and the litter usually stay inside when their mother goes hunting — an exception, probably, being the hedgehog. As for feminine nouns *medveditsa* 'female bear', *volchitsa* 'female wolf' registered in dictionaries as independent entries, it may be explained by the utility principle which is reflected in the way human empirical experience is represented in language and, consequently, in dictionaries as written artifacts for storing knowledge. On the one hand, if a hunter encounters a litter of bear or wolf cubs, he is most likely to expect to see their mother rather than the father, who cannot be far. On the other hand, an encounter with a mother bear is more dangerous as she tends to be more aggressive in protecting her cubs. So, gender distinction between large dangerous predators is clearly pragmatic and undoubtedly informative in the sense of producing an orientational effect.

8.5. Summary

The problem of (pro)nominal gender as a grammatical category is by no means exhausted by the sketchy analysis presented in this chapter. My goal was less ambitious — to show the feasibility of approaching it from a somewhat different perspective than traditional grammar which fails to provide helpful insights on its nature and function, and view it as a cognitive structure borne of human experience categorized in linguistic form. The data discussed above is strongly supportive of the view on grammar as cognition in that it illustrates the role of the observer in linguistic categorization of *perceptual experience*. Unlike traditional accounts, the suggested approach allows to make sense of otherwise incomprehensible grammatical phenomena such as the category of gender in Russian.

It has been shown that, far from being consistent with the logical principles of scientific taxonomies, gender classification of nouns in Russian reflects the results of conceptualization of the perceivable world by the languaging observer in his attempts to bring into it order and discreteness. In doing so the observer, obviously, follows some simple principle in comprehending the world, relying on his sensorium and the nature and character of his interactions with the environment. It may well be described as the WYSIWYG principle ("What You See Is What You Get"), since the world is partitioned by man not according to some 'objective' principles of its organization which are independent of the human cognizer, but as a result of human phenomenological experience. It is because of this that nominal gender as a grammatical category does not stand on its own among other grammatical categories, it is closely related to gender distinctions categorized in the system of personal pronouns and is, in fact, driven by them as far as word-formation processes go. Counter to the orthodox view on the function of pronouns in language, a biocognitive approach allows to see them as primary orientational devices in the cognitive domain of linguistic interactions; they are at the phenomenological core of language as a patterned system of coordinated orientational activity, and as such they appear to set certain paths for the cognitive processes of conceptualization and categorization reflected in other grammatical categories as metasigns grounded in perceptual experience.

BIBLIOGRAPHY

Aarts, B. (2007). *Syntactic Gradience: The nature of grammatical indeterminacy.* Oxford: Oxford University Press.

Aitchison, J. (1987). *Words in the Mind: An introduction to the mental lexicon.* Oxford UK, Cambridge USA: Blackwell.

Ajjanagadde, V. and L. Shastri (1991). Rules and variables in neural nets. *Neural Computation* 3. 121-134.

Akmajian, A., R. A. Demers, A. K. Farmer and R. M. Harnish (1990). *Linguistics: An introduction to language and communication.* Cambridge, MA: The MIT Press.

Albertazzi, L. (ed.) (2000). *Meaning and Cognition.* Amsterdam, Philadelphia: John Benjamins.

Allerton, D. J. (1990). Language as form and pattern: grammar and its categories. In N. E. Collinge (ed.), *An Encyclopaedia of Language.* London and New York: Routledge. 68-111.

Allwood, J. (1999). Semantics as meaning determination with semantic-epistemic operations. In J. Allwood and P. Gärdenfors (eds.), *Cognitive Semantics: Meaning and cognition.* Amsterdam, Philadelphia: John Benjamins. 1-17.

Allwood, J. and P. Gärdenfors (eds.) (1999). *Cognitive Semantics: Meaning and cognition.* Amsterdam, Philadelphia: John Benjamins.

Altmann, G. T. M. (1998). Ambiguity in Sentence Processing. *Trends in Cognitive Sciences* 2(4). 146-152.

Anderson, S. R. and D. W. Lightfoot (2002). *The Language Organ: Linguistics as cognitive physiology.* Cambridge: Cambridge University Press.

Austin, J. L. (1962). *How to Do Things with Words.* The William James Lectures delivered in Harvard University in 1955. Edited by J. O. Urmson. Oxford: Clarendon Press.

Bache, C. (1985). *Verbal Aspect: A general theory and its application to present-day English.* Odense: Odense University Press.

Barber, B. (1961). Resistance by scientists to scientific discovery. *Science* 84. 596-602.

Barbieri, M. (2006). Life and semiosis: The real nature of information and meaning. *Semiotica* 158(1/4). 233-254.

Barlow, M. and S. Kemmer (2000). *Usage-Based Models of Language.* Stanford, CA: CSLI Publications.

Bates, E. and J. Goodman (1998). Grammar from the lexicon. In B. MacWhinney (ed.), *The Emergence of Language.* Hillsdale, NJ: Lawrence Erlbaum. 197–212.

Bennett, M. (1978). Demonstratives and indexicals in Montague grammar. *Synthese* 39(1). 1-80.

Bennett, M.R. and P. M. S. Hacker (2002). *Philosophical Foundations of Neuroscience*. Oxford: Blackwell.
Benveniste, E. (1971). *Problems in General Linguistics*. Coral Gables, Florida: University of Miami Press.
Bickerton, D. (1981). *Roots of Language*. Ann Arbor: Karoma.
Bickerton, D. (1990). *Language and Species*. Chicago and London: The University of Chicago Press.
Bod, R. (1998). *Beyond Grammar: An experience-based theory of language*. Stanford, CA: Center for the Study of Language and Information.
Boden, M. A. (2000). Autopoiesis and life. *Cognitive Science Quarterly* 1. 117-145.
Boeckx, C. and M. Piattelli-Palmarini (2005). Language as a natural object – linguistics as a natural science. *The Linguistic Review* 22. 447-466.
Bondarko, A. V. and L. L. Bulanin (1967). *Russkij glagol* [The Russian Verb]. Leningrad: Prosveshchenije.
Borkovskij, V. I. and P. S. Kuznetsov (1963). *Istoricheskaja grammatika russkogo jazyka* [*A Historical Grammar of the Russian Language*]. Moskva: Izdatel'stvo akademii nauk.
Bottineau, D. (in press). Language and enaction. In J. Stewart, O. Gapenne and E. Di Paolo (eds), *Enaction: Towards a new paradigm for cognitive science*. Cambridge, Mass.: MIT Press.
Bouissac, P. (ed.) (1998). *Encylopedia of Semiotics*. Oxford: Oxford University Press.
Brady, M. (1989/1997). *What's Worth Teaching? Selecting, organizing, and integrating knowledge*. State University of New York (1989), Books for Educators (renewed 1997).
Braga, L. S. (1999). Peirce and biology. *Semiotica* 127(1/4) *(Special Issue: Biosemiotica)*. 5-21.
Brier, S. (2006). The cybersemiotic model of communication: An evolutionary model of the threshold between semiosis and information exchange. *Semiotica* 158(1/4). 255-296.
Brisard, F. (2005). Epistemic interactions of tense and aspect in the English verb: the paradigm of the present. In K. Turewicz (ed.), *Cognitive Linguistics: A user friendly approach*. Szczecin: Wydawnictwo Naukowe Uniwersytetu Szczcinskiego. 65-63.
Brown, R. O. (1973). *A First Language: The early stages*. Cambridge, MA: Harvard University Press.
Brugmann, K. (1889). *Grundriss der vergleichenden Grammatik der indogermanischen Sprachen. B. 2, H. 1: Wortbildungslehre*. Strassburg: Karl J. Trübner.

Brugmann, K. (1904). Die Demonstrativpronomen der indogermanischen Sprachen. *Abhandlungen der königlich sächsische Gesellschaft der Wissenschaften. Philosophisch-historische Klasse* 22(6). Leipzig: Teubner.

Budagov, R. A. (2000). *Jazyk i rech v krugozore cheloveka* [*Language and speech in the human mental outlook*]. Moscow: Dobrosvet.

Bühler, K. (1934). *Sprachtheorie*. Jena: Gustav Fischer.

Cameron-Faulkner, T., E. Lieven and M. Tomasello (2003). A construction based analysis of child directed speech. *Cognitive Science* 27. 843-873.

Cangelosi, A. (in press). Adaptive agent modeling of distributed language: investigations on the effects of cultural variation and internal action representations. *Language Sciences* (2007). doi:10.1016/j.langsci.2006.12.026.

Cassirer, E. (1923). *Philosophie der symbolischen Formen. Erster Teil: Die Sprache*. Berlin: Bruno Cassirer Verlag.

Chardin, T. de (1956). *Le phénomène humain*. Paris: Les Éditions du Seuil.

Chernykh, P. J. (1994). *Istoriko-etimologicheskij slovar' sovremennogo russkogo jazyka* [*Historical etymological dictionary of Modern Russian language. Vol. 1-2*]. Moskva: Russkij jazyk.

Chomsky, N. (1957). *Syntactic Structures*. The Hague: Mouton.

Chomsky, N. (1965). *Aspects of the Theory of Syntax*. Cambridge, Mass.: The MIT Press.

Chomsky, N. (1975). *Reflections on Language*. New York: Pantheon.

Chomsky, N. (1991). Linguistics and cognitive science: Problems and mysteries. In A. Kasher (ed.), *The Chomskyan Turn: Generative linguistics, philosophy, mathematics, and psychology*. Oxford: Blackwell. 26-55.

Chomsky, N. (2002). *Cartesian Linguistics: A chapter in the history of rationalist thought*, 2nd ed. Cybereditions Corporation (edited, with an introduction, by J. McGilvray).

Clark, A. (1997). *Being There: Putting brain, body and world together again*. Cambridge, MA: MIT Press.

Clark, A. and D. Chalmers (1998). The extended mind. *Analysis* 58(1). 7-19.

Clark, H. H. (1996). *Using Language*. Cambridge: Cambridge University Press.

Cole, D. (1999). I don't think so: Pinker on the mentalese monopoly. *Philosophical Psychology* 12. 283-295.

Cook-Gumperz, J. (2006). The social construction of literacy. In J. Cook-Gumperz (ed.), *The Social Construction of Literacy*. Cambridge: Cambridge University Press. 1-18.

Cornish-Bowden, A. and M. L. Cárdenas (2001). Complex networks of interactions connect genes to phenotypes. *Trends in Biochemical Science* 26. 463–465.

Cornish-Bowden, A., M. L. Cárdenas, J.-C. Letelier, J. Soto-Andrade, F. G. Abarzúa (2004). Understanding the parts in terms of the whole. *Biology of the Cell* 96. 713–717.

Cowley, S. (2002). Why brains matter: an integrational perspective on the Symbolic Species. *Language Sciences* 24(1). 73-95.
Cowley, S. (2004a). Contextualizing bodies: human infants and distributed cognition. *Language Sciences* 26(4). 565-591.
Cowley, S.J. (2004b). Simulating others: the basis of human cognition? *Language Sciences* 26(3). 273-299.
Cowley, S. (in press). The cognitive dynamics of distributed language. *Language sciences* (2007). doi:10.1016/j.langsci.2007.01.002.
Cowley, S. and N. Love (2006). Language and cognition, or, how to avoid the conduit metaphor. In A. Duszak and U. Okulska (eds.), *Bridges and Walls in Metalinguistic Discourse.* Frankfurt/Main: Peter Lang. 135-154.
Cowley, S.J., S. Moodley and A. Fiori-Cowley (2004). Grounding signs of culture: primary intersubjectivity in social semiosis. *Mind, Culture and Activity* 11(2). 109-132.
Croft, W. (1998). Linguistic evidence and mental representations. *Cognitive Linguistics* 9(2). 151-173.
Culicover, P. W. (2005). Linguistics, cognitive science, and all that jazz. *The Linguistic Review* 22. 227-248.
Dahl, Ö. (1985). *Tense and Aspect Systems.* Oxford: Basil Blackwell.
Damasio, A. (1999). *The Feeling of What Happens: Body and emotion in the making of consciousness.* New York; San Diego; London: Harcourt.
Danaher, D. (1998). Peirce's semiotic and conceptual metaphor theory. *Semiotica* 119(1/2). 171-207.
Dascal, M. (2006). Hobbes's challenge. In A. Brook (ed.), *The Prehistory of Cognitive Science.* Palgrave-McMillan. 67-96.
Davidson, D. (2001). *Subjective, Intersubjective, Objective.* Oxford: Oxford University Press.
Davis, B. and D. Sumara (1997). Cognition, complexity, and teacher education. *Harvard Educational Review* 67(1). 105–125.
Deacon, T. W. (1997). *The Symbolic Species: The co-evolution of language and the brain.* Harmondsworth: Penguin.
Deane, P. D. (1996). On Jackendoff's conceptual semantics. *Cognitive Linguistics* 7. 35-91.
Devitt, M., and K. Sterelny (1999). *Language and Reality. An introduction to the philosophy of language* (2nd edition). Cambridge, MA: The MIT Press.
Dewey, J. (1925/1958). *Experience and Nature.* La Salle, IL: Open Court.
Dijk, T. van. (2006). Introduction: discourse, interaction and cognition. *Discourse Studies* 8(1). 5-7.
Di Paolo, E. (2005). Autopoiesis, adaptivity, teleology, agency. *Phenomenology and the Cognitive Sciences* 4(4). 429-452.
Dirven, R. and M. Verspoor (eds.) (1998). *Cognitive Exploration of Language and Linguistics.* Amsterdam, Philadelphia: John Benjamins.

Donald, M. (1991). *Origins of the Modern Mind: Three stages in the evolution of culture and cognition.* Cambridge, MA: Harvard University Press.

Doroszewski, W. (1970). *Elementy leksykologii i semiotyki.* Warszawa: Panstwowe Wydawnictwo Naukowe.

Dummett, M. (1981). *Frege: Philosophy of Language* (2nd edition). Cambridge, MA: Harvard University Press.

Eco, U. (1984). *Semiotics and the Philosophy of Language.* Basingstoke etc.: Macmillan.

Eckardt, B. von (1993). *What Is Cognitive Science?* Cambridge, MA: The MIT Press.

Elman, J. (2004). An alternative view of the mental lexicon. *Trends in Cognitive Sciences* 8(7). 301-306.

Elman, J., E. Bates, M. Johnson, A. Karmiloff-Smith, D. Parisi and K. Plunkett (1996). *Rethinking Innateness: A connectionist perspective on development.* Cambridge, MA: MIT Press.

Emmeche, C. (1998). Defining life as a semiotic phenomenon. *Cybernetics and Human Knowing* 5(1). 3-17.

Emmeche, C. (1999). The biosemiotics of emergent properties in a pluralist ontology. In E. Taborsky (ed.), *Semiosis. Evolution. Energy: Towards a reconceptualization of the sign.* Aachen: Shaker Verlag. 89-108.

Everett, D. L. (2005). Biology and language: a consideration of alternatives. *Journal of Linguistics* 41. 157–175.

Farrow, S. (2005). The language myth revisited. *Language and Communication* 25. 19–25.

Fauconnier, G. (1999). Methods and generalizations. In Th. Janssen and G. Redeker (eds.) (1999), *Cognitive Linguistics: Foundations, scope, and methodology.* Berlin, New York: Mouton de Gruyter. 95-127.

Feldman, J. (1982). Dynamic connections in neural networks. *Biological Cybernetics* 46. *27-39.*

Fillmore, C. J. (1986). "U"-semantics, second round. *Quaderni di semantica* 7(1). 49-58.

Fodor, J. A. (1975). *The Language of Thought.* New York: Crowell.

Fodor, J. A. (1987). *Psychosemantics: The problem of meaning in the philosophy of mind.* Cambridge, MA.: Bradford Books/The MIT Press.

Fodor, J. A. (1998). *Concepts: Where cognitive science went wrong.* Oxford: Clarendon Press.

Forchheimer P. (1953). *The Category of Person in Language.* Berlin: Walter de Gruyter & Co.

Freeman, W. J. (1995). *Societies of Brains: A study in the neuroscience of love and hate.* Hillsdale, NJ: Erlbaum.

Fried, M. and H. C. Boas (eds.) (2005). *Grammatical Constructions: Back to the roots.* Amsterdam, Philadelphia: John Benjamins.

Fries, P. (1983). On the status of theme in English: arguments from discourse. In J. Petöfi and E. Sozer (eds.), *Micro and Macro Connexity of Texts*. Hamburg: Helmut Buske Verlag. 116-152.

Frisina, W. G. (2002). *The Unity of Knowledge and Action: Toward a nonrepresentational theory of knowledge* (SUNY Series in Philosophy). State University of New York Press.

Fromkin, V. and R. Rodman (1998). *An Introduction to Language*. Fort Worth etc.: Harcourt Brace.

Fuchs, C. and S. Roberts (eds.) (1999). *Language Diversity and Cognitive Representations*. Amsterdam, Philadelphia: John Benjamins.

Gamkrelidze, T. V. (1988). R. O. Jakobson i problema isomorfizma mezhdu geneticheskim kodom i semioticheskimi sistemami. *Voprosy jazykoznanija* 3. 5-8 [R. O. Jakobson and the problem of isomorphism between genetic code and semiotic systems].

Gardner, H. (1985). *The Mind's New Science: A history of the cognitive revolution*. New York: Basic Books.

Garfield, J. L. (1997). Mentalese not spoken here: Computation, cognition and causation. *Philosophical Psychology* 10. 413-435.

Gasparov, B. M. (1996). *Jazyk, pamjat, obraz. Lingvistika jazykovogo sushchestvovanija*. Moskva: Novoje Literaturnoje Obozrenije [*Language, Memory, Image: The linguistics of language being*].

Gauker, C. (2003). *Words Without Meaning*. Cambridge, MA: The MIT Press.

Gärdenfors, P. (1999). Some tenets of cognitive semantics. In J. Allwood and P. Gärdenfors (eds.), *Cognitive Semantics: Meaning and cognition*. Amsterdam, Philadelphia: John Benjamins. 19-36.

Geach, T. (1962). *Reference and Generality*. Ithaca, N. Y.: Cornell University Press.

Geeraerts, D. (1988). Cognitive grammar and the history of lexical semantics. In B. Rudzka-Ostyn (ed.), *Topics in Cognitive Linguistics*. Amsterdam/ Philadelphia: John Benjamins. 647-677.

Geeraerts, D. (1989). Prospects and problems of prototype theory. *Linguistics* 27(7). 587-612.

Gelman, S. A. and P. Bloom (2000). Young children are sensitive to how an object was created when deciding what to name it. *Cognition* 76. 91-103.

Gibson, J. (1979). *The Ecological Approach to Visual Perception*. Houghton-Mifflin, Boston MA.

Givón, T. (1993). *English Grammar: A function-based introduction*. Vol. 1-2. Amsterdam, Philadelphia: John Benjamins.

Givón, T. (1995). *Functionalism and Grammar*. Amsterdam, Philadelphia: John Benjamins.

Givón, T. (2002). *Bio-Linguistics. The Santa Barbara Lectures*. Amsterdam, Philadelphia: John Benjamins.

Glare, P. G. W. (1982). *Oxford Latin Dictionary*. Oxford: At the Clarendon Press.
Glasersfeld, E. von (1995). *Radical Constructivism: A way of knowing and learning*. London: Falmer Press.
Głaz, A. (2002). Speakers' choices: Seeing things from a perspective. In B. Lewandowska-Tomaszczyk and K. Turewicz (eds.), *Cognitive Linguistics Today*. Frankfurt/Main: Peter Lang. 287-303.
Głaz, A. (2007). Vantage theory: A newcomer to the cognitivist scene? In M. Fabiszak (ed.), *Language and Meaning: Cognitive and functional perspectives*. Frankfurt/Main: Peter Lang. 91-112.
Glovinskaja, M. J. (1982). *Semanticheskije tipy vidovyx ptotivopostavlenij russkogo glagola* [*Semantic Types of Aspectual Oppositions of the Russian Verb*]. Moskva: Nauka.
Goldberg, A. E. (1995). *Constructions: A construction grammar approach to argument structure*. Chicago: Chicago University Press.
Goldberg, A. (2005). *Constructions at Work: The nature of generalization in language*. Oxford: Oxford University Press.
Goldberg, A. E, D. M. Casenhiser and N. Sethuraman (2004). Learning argument structure generalizations. *Cognitive Linguistics* 15(3). 289–316
Goldsmith, J. and E. Woisetschlaeger (1982). The logic of the English progressive. *Linguistic Inquiry* 13(1). 79-89.
Grice, H. P. (1989). *Studies in the Way of Words*. Harvard University Press.
Gross, A. (2005). Is evidence based linguistics the solution? Is voodoo linguistics the problem? Paper presented at the LACUS Conference, August 2005, Dartmouth. (http://languag2.home.sprynet.com/f/evidence.htm)
Gross, S. (2005). The nature of semantics: On Jackendoff's arguments. *The Linguistic Review* 22. 249-270.
Grush, R. 1997. The architecture of representation. *Philosophical Psychology* 10(1). 5-23.
Gunajev, Z. S. (1977). O vyrazhenii prostranstvennykh otnoshenij v nekotorykh dagestanskikh jazykakh [On expression of spatial relations in some of the Dagestani languages]. *Voprosy jazykoznanija* 6. 126-129.
Hagège, C. (1996). *L'homme de paroles: Contribution linguistique aux sciences humaines*. Paris: Fayard.
Haiman, J. (ed.) (1985). *Iconicity in Syntax*. Amsterdam, Philadelphia: John Benjamins.
Haldane, J. (1989). Naturalism and the problem of intentionality. *Inquiry* 32. 305-322.
Halliday, M. (1967). Notes on transitivity and theme. Part II. *Journal of Linguistics* 3. 199-245.
Halliday, M. (1985). *An Introduction to Functional Grammar*. London: Edward Arnold.

Harder, P. (1999). Partial autonomy. Ontology and methodology in cognitive linguistics. In T. Janssen and G. Redeker (eds.), *Cognitive Linguistics: Foundations, scope, and methodology*. Berlin, New York: Mouton de Gruyter. 195-222.

Harnad, S. (1990). The symbol grounding problem. *Physica* D 42. 335-346.

Harnad, S. (2003). Back to the oral tradition through skywriting at the speed of thought (http://www.interdisciplines.org/defispublicationweb/papers/6).

Harnad, S. (2005a). Distributed processes, distributed cognizers, and collaborative cognition. In I. E. Dror (ed.), *Cognitive Technologies and the Pragmatics of Cognition:* Special issue of *Pragmatics & Cognition* 13(3). 501–514.

Harnad, S. (2005b). To cognize is to categorize: Cognition is categorization. In H. Cohen and C. Lefebvre (eds.), *Handbook of Categorization in Cognitive Science*. Elsevier. 20-42.

Harnad, S. and I. Dror (2006). *Distributed Cognition*: Special issue of *Pragmatics & Cognition* 14(2). 209-213.

Harris, R. (1981). *The Language Myth*. London: Duckworth.

Harris, R. (1996). *Signs, Language and Communication: Integrational and segregational approaches*. London and New York: Routledge.

Harris, R. (2004). Integrationism, language, mind and world. *Language Sciences* 26. 727-739

Harris, R. (2005). *The Semantics of Science*. Continuum International Publishing Group Ltd.

Heine, B. (1997). *Cognitive Foundations of Grammar*. Oxford: Oxford University Press.

Heintz, C. and G. Origgi (2004). *Rethinking Interdisciplinarity: Emergent issues.* (http://www.interdisciplines.org/~/papers/11)

Henry, D. and K. Henry (1969). Koyukon locationals. *Anthropological Linguistics* 11(4). 136-142.

Heschl, A. (1990). L = C, A simple equation with astonishing consequences. *Journal of Theoretical Biology* 145. 13-40.

Hewings, M. (1999). *Advanced Grammar in Use: A self-study reference and practice book for advanced learners of English.* Cambridge: Cambridge University Press.

Hill D. (1994). *Spatial Configurations and Evidential Propositions.* Nijmegen: Max Plank Institute for Psycholinguistics.

Hinrichs, L. (2006). *Codeswitching on the Web: English and Jamaican Creole in e-mail communication*. Amsterdam, Philadelphia: John Benjamins.

Hodges, B. (in press). Good prospects: ecological and social perspectives on talking together. *Language Sciences* (2007), doi:10.1016/ j.langsci.2007.01.003.

Hodges, B. H. and R. M. Baron (1992). Values as constraints on affordances: Perceiving and acting properly. *Journal for the Theory of Social Behaviour* 22. 263-294.

Hoffmeyer, J. (1996). *Signs of Meaning in the Universe*. Bloomington: Indiana University Press.

Hoffmeyer, J. (1997). Biosemiotics: Towards a new synthesis in biology. *European Journal for Semiotic Studies* 9(2). 355-376.

Hoffmeyer, J. and C. Emmeche (1991). Code-duality and the semiotics of nature. In M. Anderson and F. Merrell (eds.). *On Semiotic Modeling*. Mouton de Gruyter. 117-166.

Huddleston, R. (1984). *Introduction to the Grammar of English*. Cambridge: Cambridge University Press.

Humboldt, V. von (1820). Über das vergleichende Sprachstudium in Beziehung auf die verschiedenen Epochen der Sprachentwicklung (Werke in 5 Bdd. Darmstadt, Wissenschaftliche Buchgesellscaft, Bd. III, 1963. 1-25.

Hurford, J. R., M. Studdert-Kennedy and C. Knight (eds.) (1998). *Approaches to the Evolution of Language*. Cambrisge: Cambridge University Press.

Hutton, C. (1993). Analysis and notation: the case for a non-realist linguistics. In R. Harré and R. Harris (eds.), *Linguistics and Philosophy: The controversial interface*. Pergamon Press.

Imoto, S. (2004a). The philosophical nature of Maturana's theory of perception. *Cybernetics and Human Knowing* 11(2). 12-20.

Imoto, S. (2004b). What is the external world?: A cognitive science perspective. *Tetsugaku (Annals of the Philosophical Society of Hokkaido University)* 40. 29-44.

Imoto, S. (2005). Nothing as plenum: Lao-Tzu's *Way* and Maturana's *Substratum*. *Cybernetics and Human Knowing* 12(4). 107-114.

Itkonen, E. (2003). *What Is Language? A study in the philosophy of linguistics*. Turku.

Ivanov, V. V. (1983). *Istoricheskaja grammatika russkogo jazyka* [*A Historical Grammar of the Russian Language*]. Moskva: Prosveshchenije.

Jackendoff, R. (1983). *Semantics and Cognition*. Cambridge, MA: The MIT Press.

Jackendoff, R. (1994). *Patterns in the Mind: Language and human nature*. Basic Books.

Jackendoff, R. (1996). *The Architecture of the Language Faculty*. Cambridge, MA: The MIT Press.

Jackendoff, R. (1997). Semantics and Cognition. In S. Lappin (ed.), *The Handbook of Contemporary Semantic Theory*. Oxford: Basil Blackwell. 539-559.

Jackendoff, R. (1999). Possible stages in the evolution of the language capacity. *Trends in Cognitive Sciences* 3(7). 272-279.

Jackendoff, R. (2002). *Foundations of Language*. Oxford: Oxford University Press.

Jakobson, R. (1971). *Selected Writings*. The Hague, etc.: Mouton. Vol. 2.

Järvilehto, T. (1998). The theory of the organism-environment system: I. Description of the theory. *Integrative Physiological and Behavioral Science* 33. 317-330.

Jenkins, L. (2000). *Biolinguistics: Exploring the biology of language*. Cambridge, MA.: Cambridge University Press.

Jespersen, O. (1929). *The Philosophy of Grammar*. London : Allen & Unwin.

Johnson, M. (1987). *The Body in The Mind.* Chicago: University of Chicago Press.

Johnson-Laird, P. N. (1983). *Towards a Cognitive Science of Language, Inference, and Consciousness.* Cambridge: Cambridge University Press.

Keenan, E. and E. Stabler (2000). *Bare Grammar*. Cambridge: Cambridge University Press.

Keller, R. (1998). *A Theory of Linguistic Signs.* Oxford: Oxford University Press.

Kellogg, R. T. (1994). *The Psychology of Writing*. Oxford: Oxford University Press.

Kertész, A. (2004). *Cognitive Semantics and Scientific Knowledge: Case studies in the cognitive science of science.* Amsterdam, Philadelphia: John Benjamins.

Kim, J. (1998). *Mind in a Physical World: An essay on the mind-body problem and mental causation*. Cambridge, MA: The MIT Press.

Kirkeby, F. (1997). Event and body-mind: An outline of a post-postmodern approach to phenomenology. *Cybernetics and Human Knowing* 4(2/3). 3-34.

Kirsh, D. and P. Maglio (1994). On distinguishing epistemic from pragmatic action. *Cognitive Science* 18. 513-49.

Klein, E. (1967). *A Comprehensive Etymological Dictionary of the English Language*. Amsterdam, etc.: Elsevier Publishing Company. Vol. 1-2.

Koshelev, A. D. (1988). O referetsial'nom podxode k izucheniju semantiki vida (opyt referentsial'nogo opisanija vidovyx znachenij). V kn.: *Referentsija i problemy tekstoobrazovanija.* Moskva: Nauka. 30-52. [On the reference-based approach to the study of the semantics of aspect].

Kravchenko, A. V. (1990). K kognitivnoj teorii vremeni i vida *Filologicheskije nauki* 6. 81-90. Moskva [Toward a cognitive theory of tense and aspect].

Kravchenko, A. V. (1992). *Voprosy teorii ukazatel'nosti: Egotsentrichnost'. Deiktichnost'. Indeksal'nost'.* Izdatel'stvo Irkutskogo universiteta [*Issues in the Theory of Indication: Egocentricity. Deixis. Indexicality*].

Kravchenko, A. V. (1993). K probleme nabludatelja kak sistemoobrazujushchego faktora v jazyke [On the problem of observer as a systemic factor in language]. *Izvestija Rossijskoj akademii nauk. Serija literatury i jazyka* 52(3). 45-56. Moskva.

Kravchenko, A. V. (1995). Glagol'nyj vid i kartina mira [Verbal aspect and the world view]. *Izvestiya Rossiiskoi akademii nauk. Serija literatury i jazyka* 54(1). 49-64. Moskva.

Kravchenko, A. V. (1996). *Jazyk i vosprijatije: Kognitivnyje aspekty jazykovoi kategorizatsii.* Izdatel'stvo Irkutskogo universiteta [*Language and Perception: Cognitive aspects of linguistic categorization*].
Kravchenko, A. V. (ed.) (1997). *The English Verb: A new grammar for everyone.* Irkutsk: Irkutsk Linguistics University Press (second edition 1999).
Kravchenko, A. V. (2000). Jestestvennonauchnyje aspekty semiosisa [The scientific aspects of semiosis]. *Voprosy jazykoznanija* 1. 3-9. Moskva.
Kravchenko, A. V. (2001a). Kognitivnaja lingvistika i novaja epistemologija. *Izvestija Akademii Nauk. Serija literatury i jazyka* 60(5). 3-13 [Cognitive linguistics and new epistemology]. Moskva.
Kravchenko, A. V. (2001b). Russian verbs of spatial orientation STAND, SIT, LIE. In E. Nemeth (ed.), *Cognition in Language Use: Selected papers from the 7th International Pragmatics Conference.* Vol. 1. Antwerp: International Pragmatics Association. 216-225.
Kravchenko, A. V. (2002a). Toward a bio-cognitive philosophy of language. *Perspectives: Journal for Interdisciplinary Work in the Humanities* 1(5) (http://www.brookes.ac.uk/~ /perspectives).
Kravchenko, A. V. (2002b). A cognitive account of tense and aspect: Resurrecting 'dead' metaphors. *Anglophonia. French Journal of English Studies* 12. 199-212.
Kravchenko, A. V. (2003a). *Sign, Meaning, Knowledge: An essay in the cognitive philosophy of language.* Frankfurt/Main: Peter Lang.
Kravchenko, A. V. (2003b). The ontology of signs as linguistic and non-linguistic entities: a cognitive perspective. *Annual Review of Cognitive Linguistics* 1. 179-191.
Kravchenko, A. V. (2004). A new cognitive framework for Russian aspect. In F. Karlsson (ed.), *Proceedings of the 20th Scandinavian Conference of Linguistics,* Helsinki, January 7-9, 2004. University of Helsinki, Department of General Linguistics, Publications No. 36 (http://www.ling.helsinki.fi/ kielitiede/20scl/ Kravchenko.pdf).
Kravchenko, A. V. (2006). Cognitive linguistics, biology of cognition, and biosemiotics: bridging the gaps. *Language Sciences* 28(1). 51-75.
Kravchenko, A. V. (2007). Gipoteza Sepira-Uorfa v kontekste biologii poznanija [The Sapir-Whorf hypothesis from the perspective of the biology of cognition]. *Voprosy kognitivnoj lingvistiki* 1. 5-14.
Kress, G. (1997). *Before Writing: Rethinking the paths to literacy.* London: Routledge.
Kubriakova, E. S. (2002). Kognitivnaja lingvistika i problemy kompozitsionalnoj semantiki v sfere slovoobrazovanija. *Izvestija Akademii nauk. Serija literatury i jazyka* 61(1). 13-24 [Cognitive linguistics and the problem of semantic blending in word formation]

Kuhn, T. S. (1962). *The Structure of Scientific Revolutions*. Chicago: The University of Chicago Press.
Kuhn, T. S. (1977). *The Essential Tension: Selected studies in scientific tradition and change*. Chicago: The University of Chicago Press.
Kull, K. (1999). Biosemiotics in the twentieth century: a view from biology. *Semiotics* 127(1/4). 385-414.
Lakoff, G. (1987). *Women, Fire, and Dangerous Things*. Chicago: The University of Chicago Press.
Lakoff, G. and M. Johnson (1999). *Philosophy in the Flesh: The embodied mind and its challenge to western thought*. Basic Books.
Lamb, S. (1999). *Pathways of the Brain: The neurocognitive basis of language*. Amsterdam, Philadelphia: John Benjamins.
Lamb, S. (2004). *Language and Reality*. London; New York: Continuum.
Lamb, S. (2006). *The Anatomy of Language*. A workshop presented at the 2nd Biennial Conference on Cognitive Science. 9-13 June, St.-Petersburg.
Langacker, R. W. (1987). *Foundations of Cognitive Grammar, Vol. 1: Theoretical prerequisites*. Stanford University Press.
Langacker, R. W. (1988). Autonomy, agreement, and cognitive grammar. In D. Bentari, G. Larson and L. MacLeod (eds.), *Papers from the 24th Annual Meeting of the Chicago Linguistic Society*. Part Two: *Parasession on Agreement in Grammatical Theory*. Chicago: Chicago Linguistic Society. 147-180.
Langacker, R. W. (1991). *Concept, Image, and Symbol: The cognitive basis of grammar*. Berlin, New York: Mouton de Gruyter.
Larson, R. and G. Segal (1995). *Knowledge of Meaning: An introduction to semantic theory*. Cambridge, MA: The MIT Press.
Lavoisier, A. L. (1789). *Traité elémentaire de chimie*. Third paragraph. Paris: Chez Cuchet.
Lawrence, H. (1972). Viewpoint and location in Oksapmin. *Anthropological Linguistics* 14(8). 311-316.
Leech, G. and J. Svartvik (1975). *A Communicative Grammar of English*. Longman Group Ltd.
Lepore, E. and Z. Pylyshyn (eds.) (1999). *What is Cognitive Science?* Basil Blackwell.
Levi-Strauss, C. (1962) *La pensee sauvage*. Paris: Librairie Plon.
Lewin, K. (1946). Behavior and development as a function of the total situation. In L. Carmichael (ed.), *Manual of Child Psychology*. New York: Wiley. 791-844.
Letelier, J. C., G. Marin and J. Mpodozis (2003). Autopoietic and (***M,R***) systems. *Journal of Theoretical Biology* 222. 261-272.
Leyton, M. (1999). New foundations for perception. In E. Lepore and Z. Pylyshyn (eds.), *What Is Cognitive Science?* Blackwell Publishers. 121-171.

Linell, P. (1998). *Approaching Dialogue: Talk, interaction and contexts in dialogical perspectives.* Amsterdam, Philadelphia: John Benjamins.

Linell, P. (2005). *The Written Language Bias in Linguistics*. Routledge.

Linell, P. (in press). Dialogical language, dialogical minds, dialogical brains. *Language Sciences* (2007), doi:10.1016/j.langsci.2007.01.001.

Locke, J. (1961). *An Essay Concerning Human Understanding*. London: Dent (original work published 1690).

López-García, Á. (2005). *The Grammar of Genes: How the genetic code resembles the linguistic code.* Peter Lang.

Loritz, D. (1999). *How the Brain Evolved Language*. Oxford University Press.

Love, N. (2004). Cognition and the language myth. *Language Sciences* 26(6). 525-544.

Love, N. (in press). Are languages digital codes? *Language Sciences* (2007), doi:10.1016/j.langsci.2007.01.008.

Lowth, R. (1762/1767) *A Short Introduction to English Grammar: With critical notes. A new edition, corrected.* London: Published for A. Millar and T. Cadell.

Ludlow, P. (1993). Do T-theories Display Senses? *The Electronic Journal of Analytic Philosophy* 1(5) (http://www.phil.indiana.edu/ejap/).

Luria, A. R. (1998). *Jazyk i soznanije.* Rostov-na-Donu: Feniks [*Language and Mind*].

Lyman, S. M. and M. B. Scott (1966). Territoriality: a neglected social dimension. *Social Problems* 15. 236-249.

Lyons, J. (1968). *Introduction to Theoretical Linguistics*. Cambridge, etc.: Cambridge University Press.

Lyons, J. (1977). *Semantics. Volume 1.* Cambridge: Cambridge University Press.

MacLaury, R. E. (1995). Vantage theory. In J. R. Taylor and R. E. MacLaury (eds.), *Language and the Cognitive Construal of the* World. Berlin; New York: Mouton de Gruyter. 231-276.

MacLaury, R. E. (2002). Introducing vantage theory. *Language Sciences* 24(5/6). 493-536.

Malinowski, B. (1927). The Problem of meaning in primitive languages. In C. K. Ogden and I. A. Richards, *The meaning of Meaning: A study of the influence of language upon thought and of the science of symbolism.* 2nd revised edition (Supplement I, pp. 296-336). New York: Harcourt, Brace and Company.

Margolis, E. and S. Laurence (eds.) (1999). *Concepts.* Cambridge, MA: The MIT Press.

Margulis, L. (1989). Gaia: the living earth: Dialogue with Fritjof Capra. *The Elmwood Newsletter, The Elmwood Newsletter* 5(2). Berkeley, Cal..

Matan, A. and S. Carey (2001). Developmental changes within the core of artifact concepts. *Cognition* 78. 1-26.

Matthiessen, C. and S. Thompson (1989). The structure of discourse and 'subordination'. In J. Haiman and S. Thompson (eds.) (1989), *Clause Combining in Grammar and Discourse.* Amsterdam, Philadelphia: John Benjamins. 275-329.
Maturana, H. R. (1970). *Biology of Cognition.* BCL Report # 9.0. University of Illinois, Urbana.
Maturana, H. R. (1975). The organization of the living: A theory of the living organization. *International Journal of Man-Machine Studies* 7. 313-332.
Maturana, H. (1978). Biology of language: The epistemology of reality. In G. Miller and E. Lenneberg (eds.), *Psychology and Biology of Language and Thought.* New York: Academic Press. 28-62.
Maturana, H. (1983). On the misuse of the notion of information in biology. *Journal of Social and Biological Structures* 6. 155-158.
Maturana, H. R. (2000). The nature of the laws of nature. *Systems Research and Behavioral Science* 17(5). 459-468.
Maturana, H. and F. Varela (1980). *Autopoiesis and Cognition: The realization of the living.* Boston: D. Reidel.
Maturana, H. and Varela, F. (1987). *The Tree of Knowledge: The biological roots of human understanding.* Boston: Shambhala.
Maturana, H., J. Mpodozis, and J. C. Letelier (1995). Brain, language, and the origin of human mental functions. *Biological Research* 28. 15-26.
Maturana, H. R., G. Uribe and S. Frenk (1968). A biological theory of relativistic colour coding in the primate retina. *Archivos de Biologia y Medicina Experimentales (Arch. Biol. M ed. Exp.)* 1. 1-30.
McGee, K. (2005a). Enactive cognitive science. Part 1: Background and research themes. *Constructivist Foundations* 1(1). 19-34.
McGee, K. (2005b). Enactive cognitive science. Part 2: Methods, insights, and potential. *Constructivist Foundations* 1(2). 73-82.
Menary, R. (in press). Writing as thinking. *Language Sciences* (2007), doi:10.1016/j.langsci.2007.01.005.
Mey, M. de (1982). *The Cognitive Paradigm: An integrated understanding of scientific development.* Holland: D. Reidel.
Miller, G. A. and P. N. Johnson-Laird. *Language and Perception.* Cambridge, MA: The Belknap Press of Harvard University Press.
Miller, G. A. and E. Lenneberg (eds.) (1978). *Psychology and Biology of Language and Thought.* New York: Academic Press. 28-62.
Millikan, R. (1984). *Language, Thought and Other Biological Categories.* Cambridge, MA: The MIT Press.
Millikan, R. G. (2005). *Language: A biological model.* Oxford: Clarendon Press.
Miloslavskij, I. G. (1989). Vid russkogo glagola kak slovoobrazovatel'naja kategorija [Russian verbal aspect as a word-formation category]. *Filologicheskije nauki* 4. 37-44. Moskva.

Moreno, A., J. J. Merelo A. Etxeberria (1992). Perception, adaptation and learning. In B. McMullin and N. Murphy (eds.), *Autopoiesis and Perception: A workshop with ESPRIT BRA 3352.* Dublin. 65-70.
Morris, C. W. (1938/1945). Foundations of the theory of signs. In O. Neurath, R. Carnap and C. W. Morris (eds.), *International Encyclopedia of Unified Science, vol. 1, part 2.* Chicago.
Morris, C. W. (1971). *Writings on the General Theory of Signs* (ed. by T. A. Sebeok). The Hague-Paris: Mouton.
Murphy, N. (1992). The causal and symbolic explanatory duality as a framework for understanding vision. In B. McMullin and N. Murphy (eds.), *Autopoiesis and Perception: A workshop with ESPRIT BRA 3352.* Dublin. 15-25.
Murphy, R. (1985). *English Grammar in Use: A self-study reference and practice book for intermediate students.* Cambridge: Cambridge University Press.
Narayanan, S. (1997). *Embodiment in Language Understanding: Sensory-motor representations for metaphoric reasoning about event descriptions.* Ph.D. Dissertation. Department of Computer Science, University of California, Berkeley.
Nelson, D. G. K., A. Frankenfield, C. Morris and E. Blair (2000). Young children's use of functional information to categorize artifacts: three factors that matter. *Cognition* 77. 133-168.
Newell, A. (1990). *Unified Theories of Cognition.* Cambridge, MA: Harvard UP.
Ogden, C. K. and I. A. Richards (1923/1927). *The Meaning of Meaning: A study of the influence of language on thought and of the science of symbolism.* 2nd revised edition. New York: Harcourt, Brace and Company.
Oleksy, M. (2002). Cognition, language, and praxis: Is cognitive linguistics on the verge of the practical turn? In B. Lewandowska-Tomaszczyk and K. Turewicz (eds.), *Cognitive Linguistics Today.* Frankfurt/Main: Peter Lang. 55-64.
Olson, D. R. (1994). *The World on Paper: The conceptual and cognitive implications of writing and reading.* Cambridge: Cambridge University Press.
Olson, D. R. and N. Torrance (eds.) (2001). *The Making of Literate Societies.* Blackwell Publishers.
Õim, H. Towards a theory of linguistic pragmatics. *Journal of Pragmatics* 1(3). 251-267.
Paducheva E. V. (1986). Semantika vida i tochka otscheta [Semantics of aspect and the point-of-reference]. *Izvestija akademii nauk SSSR. Serija literatury i jazyka* 45(5). 413-424.
Pajunen, A. (ed.) (2002). *Mimesis, Sign and Language Evolution* (Publications in General Linguistics 3). University of Turku.
Palmer, F. R. (1968). *A Linguistic Study of the English Verb.* Coral Gables, Florida: University of Miami Press.
Palmeri, T. J. and C. Blalock (2000). The role of background knowledge in speeded perceptual categorization. *Cognition* 77. B45-B57.

Paradis, C. (2003). Is the notion of *linguistic competence* relevant in cognitive linguistics? *Annual Review of Cognitive Linguistics* 1. 207-231.
Pateman, T. (1987). *Language in Mind and Language in Society: Studies in linguistic reproduction*. Oxford: Clarendon Press.
Peirce, C. S. (1932). Elements of Logic. In C. Hartshorne and P. Weiss (eds.), *Collected Papers of Charles Sanders Peirce, Vol. 2*. Harvard University Press: Cambridge MA.
Peirce, C. S. (1958). *Logical Foundations of the Theory of Signs*. Boston: Harvard University Press.
Petrilli, S. (1999). Charles Morris's biocemiotics. *Semiotica* 127(1/4) *(Special Issue: Biosemiotica)*. 67-102.
Pettersson, J. S. (1996). *Grammatological Studies: Writing and its relation to speech*. Uppsala: Department of Linguistics.
Philipse, H. (1992). Heidegger's question of being and the Augustinian picture of language. *Philosophy and Phenomenological Research* 52(2). 251-287.
Pinker, S. (1995a). Facts about human language relevant to its evolution. In J.-P. Changeux and J. Chavaillon (eds.), *Origins of the Human Brain*. Oxford: Clarendon Press. 262-283.
Pinker, S. (1995b). *The Language Instinct: How the mind creates language*. New York, NY: Harper Perennial.
Pinker, S. (1999). *Words and Rules*. New York: Basic Books.
Pinker, S. and R. Jackendoff (2005). The faculty of language: what's special about it? *Cognition* 95. 201-236.
Platts, M. (1997). *Ways of Meaning: An introduction to a philosophy of language* (2nd edition). Cambridge, MA: The MIT Press.
Plotkin, H. (1987). Evolutionary epistemology and the synthesis of biological and social science. In W. Callebaut and R. Pinxten (eds.), *Evolutionary Epistemology: A multiparadigm program*. D. Reidel. 75-96.
Pollack, R. (1994). *Signs of Life: The language and meaning of DNA*. London: Viking.
Postal, P. M. (2004). *Skeptical Linguistic Essays*. Oxford: Oxford University Press.
Potebnja A. A. (1977). *Iz zapisok po russkoj grammatike* [From the Notes on Russian Grammar]. Vol. 4. Moskva: Prosveshchenije.
Premack, D. (1990). The infant's theory of self-propelled objects. *Cognition* 36(1). 1-16.
Priest, S. (1991). *Theories of the Mind: A compelling investigation into the ideas of leading philosophers on the nature of the mind and its relation to the body*. Harmondsworth: The Penguin Books.
Prior, A. (1967). *Past, Present and Future*. Oxford: Clarendon Press.
Prior, A. and K. Fine (1977). *Worlds, Times and Events*. Amherst: University of Massachusetts Press.

Putnam, H. (1983). *Philosophical Papers.* Vol. 3: *Realism and Reason.* Cambridge University Press.
Putnam, H. (1988). *Representation and Reality.* Cambridge MA: The MIT Press.
Pylyshyn, Z. (1999). What's in your mind? In E. Lepore and Z. Pylyshyn (eds.), *What is Cognitive Science?* Blackwell Publishers. 1-25.
Quine, W. V. (1961). Logic as a source of syntactical insight. In R. Jakobson (ed.), *Structure of Language and its Mathematical Aspects.* Providence: American Mathematical Society.
Quine, W. V. (1964). Time. In J. J. C. Smart (ed.), *Problems of Space and Time.* New York etc.: Macmillan. 370-374.
Quirk, R., S. Greenbaum, G. Leech and J. Svartvik (1973). *A University Grammar of English.* Longman Group Ltd.
Radden, G. and K-U. Panther (eds.) (2004). *Studies in Linguistic Motivation.* Berlin; New York: Mouton de Gruyter.
Reddy, M. (1979). The conduit metaphor. In A. Ortony (ed.), *Metaphor and Thought.* Cambridge: Cambridge University Press. 284-324.
Reed, E.S. (1991). Cognition as the cooperative appropriation of affordances. *Ecological Psychology* 3. 135–158.
Reed, E.S. (1996). *Encountering the World: Toward an ecological psychology.* Oxford: Oxford University Press.
Reeke, G. N. and G. M. Edelman (1988). Real brains and artificial intelligence. *Deadalus* 117. 143-174.
Regier, T. (1995). A model of the human capacity for categorizing spatial relations. *Cognitive Linguistics* 6(1). 63-88.
Resnick, L. B. (1987). *Education and Learning to Think.* Washington, D.C.: National Academic Press.
Ritter, N. A. (2005). On the status of linguistics as a cognitive science. *The Linguistic Review* 22. 117-133.
Rosch, E. (1999). Reclaiming concepts. In R. Núñez and W. J. Freeman (eds.), *Reclaiming Cognition: The primacy of action, intention, and emotion.* Thorverton, UK: Imprint Academic. 61-77.
Ross, D. (2004). Metalinguistic signaling for coordination amongst social agents. *Language Sciences* 26(6). 621-642.
Ross, D. (in press). *H. sapiens* as ecologically special: what does language contribute? *Language Sciences* (2007), doi:10.1016/j.langsci.2006.12.008.
Ross, D. and Spurrett, S. (2004). What to say to a sceptical metaphysician: a defence manual for cognitive and behavioural scientists. *Behavioral and Brain Sciences* 27. 603-647.
Rothschild, F. (1962). Laws of symbolic mediation in the dynamics of self and personality. *Annals of New York Academy of Sciences* 96. 774-784.

Rumelhart, D. E. (1979). Some problems with the notion that words have literal meanings. In A. Ortony (ed.), *Metaphor and Thought*. Cambridge: Cambridge University Press. 71-82.
Russell, B. (1948). *Human Knowledge: Its scope and limits.* New York: Simon and Schuster.
Ryle, G. (1949). *The Concept of Mind.* London: Hutchinson.
Salmon, N. (1986). *Frege's Puzzle*. Cambridge, MA: The MIT Press.
Sampson, G. (1997). *Educating Eve: The "Language Instinct" debate*. New York: Cassell.
Sandra, D. (1998). What linguistics can and can't tell you about the human mind: A reply to Croft. *Cognitive Linguistics* 9. 361-378.
Sapir, E. (1929). The status of linguistics as a science. *Language* 5. 207-214.
Saussure, F. de (1916/1922). *Cours de linguistique générale*. Paris: Payot and Cie.
Schlechtman, M. (1997). The brain/body problem. *Philosophical Psychology* 10(2). 149-164.
Schult, J. (1989). Biosemiotik — Gegenstandsbereiche und Anwendungsmöglichkeiten: Anmerkungen zum Verhältnis von Biologie und Semiotik. *Kodikas/Code* 12(3/4). 261-274.
Schyns, P. G. (1997). Categories and percepts: a bi-directional framework for categorization. *Trends in Cognitive Sciences* 1. 183-189.
Schrödinger, E. (1959). *Mind and Matter. The Tarner lectures delivered at Trinity College, Cambridge, in October 1956.* Cambridge: At the University Press.
Searle, J. (1984). *Minds, Brains and Science.* Cambridge, Mass.: Harvard University Press.
Searle, J. (1998). *Mind, Language and Society: Philosophy in the real world.* New York: Basic Books.
Sebeok, T. (1963). Communication in animals and men. *Language* 39. 448-466.
Sebeok, T. (1979). *The Sign and Its Masters.* University of Texas Press.
Sebeok, T. (1994). *Signs: An Introduction to Semiotics*. Toronto: University of Toronto Press.
Sebeok, T. A. (2001). Biosemiotics: its roots, proliferation and prospects. *Semiotica* 134(1/4). 61-78.
Sebeok, T. and J. Umiker-Sebeok (eds.) (1992). *Biosemiotics: The Semiotic Web 1991*. Berlin; New York: Mouton de Gruyter.
Sebeok, T., J. Hoffmeyer and C. Emmeche (eds.) (1999). *Biosemiotica* (Special Issue of *Semiotica* 127(1/4)). Berlin, New York: Mouton de Gruyter.
Seuren, P. A. (1985). *Discourse Semantics*. Oxford: Blackwell Publishers.
Seuren, P. A. (1998). *Western Linguistics: An historical introduction*. Oxford: Blackwell Publishers.
Shannon, C. (1948). A mathematical theory of communication. *The Bell System Technical Journal* 27. 379–423, 623–656.

Sharov, A. A. (1992). Biosemiotics: Functional-evolutionary approach to the analysis of the sense of information. In T. Sebeok and J. Umiker- Sebeok (eds.), *Biosemiotics: The Semiotic Web 1991*. Berlin; New York: Mouton de Gruyter. 345-373.

Sharov, A. A. (1998). From cybernetics to semiotics in biology. *Semiotica* 120(3/4). 403-419.

Shaumjan, S. (2001). O ponjatii jazykovogo znaka [On the notion of linguistic sign]. E. S. Kubriakova, T. J. Janko (eds.), *Jazyk i kul'tura: Fakty i tsennosti.* Moskva.: Jazyki slavjanskoj kul'tury. 149 – 165.

Sherer, V. E. (1979). O vyrazhenii prostranstvennykh otnoshenij v ketskom jazyke [On the expression of spatial relations in the Ket language]. *Voprosy jazykoznanija* 5. 103-108.

Shibatani, M. and S. Thompson (eds.) (1996). *Grammatical Constructions: Their form and meaning.* Oxford: Oxford University Press.

Shvedova, N. J. (ed.) (1980). *Russkaja grammatika* [*Russian Grammar*, in 2 volumes]. Moskva: Nauka.

Shvedova, N. J. and V. V. Lopatin (1989). *Kratkaja russkaja grammatika* [*An Abridged Grammar of Russian*]. Moskva: Russkij jazyk.

Sinha, C. (1999). Grounding, mapping, and acts of meaning. In T. Janssen and G. Redeker (eds.), *Cognitive Linguistics: Foundations, scope, and methodology.* Berlin, New York: Mouton de Gruyter. 223-255.

Skeat, W. (1956). *An Etymological Dictionary of the English Language.* Oxford: Clarendon Press.

Slane, S., R. Petruska and S. Cheyfitz (1981). Personal space measurement: a validational comparison. *Psychological Record* 31(2). 145-151.

Smith, A. (2003). Through the interdisciplinary looking glass: The rhetoric of curriculum change. *Perspectives: Journal for Interdisciplinary Work in the Humanities* 1(6) (http://www.brookes.ac.uk/~ /perspectives).

Sommer, R. (1969). *Personal Space: The behavioral basis of design.* Englewood Cliffs, N. J.: Prentice-Hall.

Spurrett, D. (2004). Distributed cognition and integrational linguistics. *Language Sciences* 26(4). 497–501.

Stepanov, J. S. (1971). *Semiotika*. Moskva: Nauka.

Stepanov, J. S. (1976). Vid, zalog, perexodnost (Balto-slavjanskaja problema I) [Aspect, voice, transitivity (The Balto-Slavic problem I)]. *Izvestija akademii nauk SSSR. Serija literatury i jazyka* 35(5). 408-420.

Stepanov, J. S. (1981). *Imena. Predikaty. Predlozhenija (Semiologicheskaja grammatika).* Moskva: Nauka [*Names. Predicates. Sentences (A Semiological Grammar)*].

Stich, S. (1992). What is a theory of mental representation? *Mind* 101. 243-261.

Stoevsky, A. (2001). Aspect, perception and markedness. Proceedings of the conference *Perspectives on Aspect.* December 12 - 14, Utrecht. http://www-Uilots.let.uu.nl/conferences/Perspectives_on_Aspect/P_o_A_procpapers.html
Strømnes, F. J. (2006). *The Fall of the Word and the Rise of the Mental Model: A reinterpretation of the recent research on spatial cognition and language.* Frankfurt/Main etc.: Peter Lang.
Sutton, J. (2004). Representation, levels, and context in integrational linguistics and distributed cognition. *Language Sciences* 26(6). 503-524.
Svorou, S. (1994). *The Grammar of Space.* Amsterdam, Philadelphia: John Benjamins.
Tarski, A. (1956). *Logic, Semantics, Metamathematics.* Oxford: Clarendon Press.
Taschek, W. W. (1992). Frege's puzzle, sense, and information content. *Mind* 101(404). 767-791.
Taylor, J. R. (1989). *Linguistic Categorization: Prototypes in linguistic theory.* Oxford: Clarendon Press.
Taylor, T. (1992). *Mutual Misunderstanding.* Durham: Duke University Press.
Taylor, T. (1997). *Theorizing Language.* Amsterdam: Pergamon Press.
Taylor, J. and R. MacLaury (eds.) (1995) *Language and the Cognitive Construal of the World.* Berlin: Mouton de Gruyter.
Thavenius, C. (1983). *Referential Pronouns in English Conversation.* CWK Gleerup.
Thibault, P. J. (2000). The dialogical integration of the brain in social semiosis: Edelman and the case for downward causation. *Mind, Culture and Activity* 7. 291-311.
Tomasello, M. (1999). *The Cultural Origins of Human Cognition.* Harvard University Press.
Tomasello, M. (2000). First steps toward a usage-based theory of language acquisition. *Cognitive Linguistics* 11. 61-82.
Tomasello, M. (2003). *Constructing a Language: A usage-based theory of language acquisition.* Harvard University Press.
Tomasello, M. (2005). Beyond formalities: The case of language acquisition. *The Linguistic Review* 22. 183-197.
Tomasello, M. and H. Rakoczy (2003). What makes human cognition unique? From individual to shared to collective intentionality. *Mind and Language* 18(2). 121-147.
Tomasello, M., M. Carpenter, J. Call, T. Behne and H. Moll (2005). Understanding and sharing intentions: The origins of cultural cognition. *Behavioral and Brain Sciences* 28(5). 675-735.
Tomlin R. (ed.) (1987). *Coherence and Grounding in Discourse.* Amsterdam, Philadelphia: John Benjamins.
Trask, R. L. (1993). *A Dictionary of Grammatical Terms in Linguistics.* Routledge.

Turewicz, K. (2000). *Applicability of Cognitive Grammar as a Foundation of Pedagogical/Reference Grammar.* Lódz: Wydawnictwo Uniwersytetu Lódzkiego.
Uexküll, J. von (1909). *Umwelt und Innenwelt der Tieren.* Berlin: J. Springer.
Vaina, L. M. (1990). 'What' and 'where' in the human visual system: two hierarchies of visual modules. *Synthese* 83(1). 49-91.
Van Valin, R. D. and R. L. LaPolla (1997). *Syntax: Structure, meaning and function.* Cambridge: Cambridge University Press.
Varela, F. J. (1992). Autopoiesis and a biology of intentionality. In B. McMullin and N. Murphy (eds.), *Autopoiesis and Perception: A workshop with ESPRIT BRA* 3352. Dublin. 4-14.
Varela, F. J., E. Thompson and E. Rosch (1991). *The Embodied Mind: Cognitive science and human experience.* Cambridge, MA: The MIT Press.
Vehkavaara, T. (1998). Extended concept of knowledge for evolutionary epistemology and for biosemiotics. In G. L. Farre and T. Oksala (eds.), *Emergence, Complexity, Hierarchy, Organization: Selected and edited papers from ECHO III.* [*Acta Polytechnica Scandinavica: Mathematics, Computing and Management in Engineering Series 91*]. Espoo. 207-216.
Vendler, Z. (1967). *Linguistics in Philosophy.* Ithaca, N.Y.: Cornell University Press.
Vendler, Z. (1968). *Activities and Nominalizations.* The Hague: Mouton.
Verbrugge, R. R. (1985). Language and event perception: Steps toward a synthesis. In W. H. Warren, Jr. and R. E. Shaw (eds.), *Persistence and Change*. Hillsdale, NJ: Lawrence Erlbaum Associates. 157-194.
Verhagen, A. (2005). *Construction of Intersubjectivity: Discourse, syntax, and cognition.* Oxford University Press.
Verschueren, J. (1999), *Understanding Pragmatics.* London, etc.: Arnold.
Viger, C. (2007). The acquired language of thought hypothesis: A theory of symbolic grounding. In T. Belpaeme, S. Cowley and K. F. MacDorman (eds.), *Symbol Grounding*: Special issue of *Interaction Studies* 8(1). 125–142.
Vygotsky, L. S. (1987). Thinking and speech. In R. W. Rieber and A. S. Carton (eds.), *The Collected Works of L. S. Vygotsky, Volume 1: Problems of General Psychology*. New York: Plenum. 39-285.
Walenski, M. and M. T. Ullman (2005). The science of language. *The Linguistic Review* 22. 327-346.
Wallis, G. and H. Bülthoff (1999). Learning to recognize objects. *Trends in Cognitive Sciences* 3. 22-31.
Watson, R. A. (1995). *Representational Ideas: From Plato to Patricia Churchland.* Dordrecht: Kluwer Academic.
Weber, A. and F. J. Varela (2002). Life after Kant: Natural purposes and the autopoietic foundations of biological individuality. *Phenomenology and the Cognitive Sciences* 1. 97-125.

Wells, G. (2000). From action to writing: Modes of representing and knowing. In J. W. Astington (ed.) *Minds in the Making: Essays in honor of David R. Olson.* Oxford: Blackwell Publishers. 115-140.

Willett, T. (1987). A cross-linguistic survey of the grammaticization of evidentiality. *Buffalo Working Papers in Linguistics* 87. 123-79.

Wheeler, M. (2004). Is language the ultimate artifact? *Language Sciences* 26(6). 693-715.

Whorf, B. L. (1956). *Language, Thought and Reality: Selected writings by Benjamin Lee Whorf.* Cambridge, MA: The MIT Press.

Wierzbicka, A. (1996). *Semantics: Primes and universals.* Oxford; New York: Oxford University Press.

Wilkins, W. K. (2005). Anatomy matters. *The Linguistic Review* 22. 271-288.

Wilson, D. (1997). Linguistic structure and inferential communication. Paper presented at the 16th International Congress of Linguists. Paris, July 25.

Wittgenstein, L. (1921/1961). *Tractatus Logico-Philosophicus.* London: Routledge and Kegan Paul.

Xrakovskij V. S. (1990). Vzaimodejstvije grammaticheskix kategorij glagola [Interplay of grammatical categories of the verb]. *Voprosy jazykoznanija* 5. 18-36. Moskva.

Yates, F. A. (1934). *John Florio: The Life of an Italian in Shakespeare's England.* Cambridge: Cambridge University Press.

Zhinkin, N. I. (1998). *Jazyk. Rech'. Tvorchestvo.* Moskva: Labirint [Language. Speech. Creativity].

Zlatev, J. (1997). *Situated Embodiment: Studies in the emergence of spatial meaning.* Stockholm: Gotab.

Zlatev, J. (2003). Meaning = life (+ culture): an outline of a unified biocultural theory of meaning. *Evolution of Communication* 4(2). 253-296.

Zlatev, J., T. Persson and P. Gärdenfors (2005). Bodily mimesis as "the missing link" in human cognitive evolution. *Lund University Cognitive Studies* 121. 1-40.

Zvegintsev, V. A. (1996). *Mysli o lingvistike* [*Thoughts about Linguistics*]. Izdatel'stvo Moskovskogo Universiteta.

Žarnić, B. (2002). Learning to learn: an epistemological paradox in education. *Synthesis Philosophica* 16(2). 355-362.

www.ingramcontent.com/pod-product-compliance
Ingram Content Group UK Ltd.
Pitfield, Milton Keynes, MK11 3LW, UK
UKHW020122290926
15471UKWH00003B/215